An Ugly Truth

An Ugly Truth

Inside
Facebook's
Battle for
Domination

Sheera Frenkel and **Cecilia Kang**

HARPER
An Imprint of HarperCollins*Publishers*

HarperCollins books may be purchased for educational, business, or sales promotional use. For information, please email the Special Markets Department at SPsales@harpercollins.com.

FIRST EDITION

Designed by Bonni Leon-Berman

Library of Congress Cataloging-in-Publication Data has been applied for.

ISBN 978-0-06-296067-2

21 22 23 24 25   LSC   10 9 8 7 6 5 4 3 2 1

To Tigin, Leyla, Oltac, 엄마, 아빠
To Tom, Ella, Eden, אמא, אבא

Contents

Authors' Note

This book is the product of more than a thousand hours of interviews with more than four hundred people, the majority of whom are executives; former and current employees and their families, friends, and classmates; and investors in and advisers of Facebook. We also drew on interviews with more than one hundred lawmakers and regulators and their aides; consumer and privacy advocates; and academics in the United States, Europe, the Middle East, South America, and Asia. The people interviewed participated directly in the events described or, in a few instances, were briefed on the events by people directly involved. Mentions of *New York Times* reporters in certain scenes refer to us and/or our colleagues.

An Ugly Truth draws on never-reported emails, memos, and white papers involving or approved by top executives. Many of the people interviewed recalled conversations in great detail and provided contemporaneous notes, calendars, and other documents we used to reconstruct and verify events. Because of ongoing federal and state litigation against Facebook, nondisclosure agreements in employment contracts, and fears of reprisal, the majority of interviewees spoke on the condition of being identified as a source rather than by name. In most cases, multiple people confirmed a scene, including eyewitnesses or people briefed on the episode. Therefore, readers should not assume the individual speaking in a given scene provided that information. In instances where Facebook spokespeople denied certain events or characterizations of its leaders and scenes, multiple people with direct knowledge verified our reporting.

The people who spoke to us, often putting their careers at risk, were crucial to our ability to write this book. Without their voices, the story of the most consequential social experiment of our times could not have been told in full. These people provide a rare look inside a company whose stated mission is to create a connected world of open expression, but whose corporate culture demands secrecy and unqualified loyalty.

While Zuckerberg and Sandberg initially told their communications staff that they wanted to make sure their perspectives were conveyed in this book, they refused repeated requests for interviews. On three occasions, Sandberg invited us to off-the-record conversations in Menlo Park and New York, with the promise that those conversations would lead to longer interviews for the record. When she learned about the critical nature of some of our reporting, she cut off direct communication. Apparently the unvarnished account of the Facebook story did not align with her vision of the company and her role as its second-in-command.

Zuckerberg, we were told, had no interest in participating.

An Ugly Truth

At Any Cost

Mark Zuckerberg's three greatest fears, according to a former senior Facebook executive, were that the site would be hacked, that his employees would be physically hurt, and that regulators would one day break up his social network.

At 2:30 p.m. on December 9, 2020, that last fear became an imminent threat. The Federal Trade Commission and nearly every state in the nation sued Facebook for harming its users and competitors, and sought to dismantle the company.

Breaking news alerts flashed across the screens of tens of millions of smartphones. CNN and CNBC cut from regular programming to the announcement. The *Wall Street Journal* and the *New York Times* posted banner headlines across the tops of their home pages.

Minutes later, New York State Attorney General Letitia James, whose office coordinated the bipartisan coalition of forty-eight attorneys general, held a press conference in which she laid out the case, the strongest government offensive against a company since the breakup of AT&T in 1984. What she claimed amounted to a sweeping indictment of Facebook's entire history—and specifically of its leaders, Mark Zuckerberg and Sheryl Sandberg.

"It tells a story from the beginning, the creation of Facebook at

Harvard University," James said. For years, Facebook had exercised a merciless "buy-or-bury" strategy to kill off competitors. The result was the creation of a powerful monopoly that wreaked broad damage. It abused the privacy of its users and spurred an epidemic of toxic and harmful content reaching three billion people. "By using its vast troves of data and money, Facebook has squashed or hindered what the company perceived as potential threats," James said. "They've reduced choices for consumers, they stifled innovation and they degraded privacy protections for millions of Americans."

Cited more than one hundred times by name in the complaints, Mark Zuckerberg was portrayed as a rule-breaking founder who achieved success through bullying and deception. "If you stepped into Facebook's turf or resisted pressure to sell, Zuckerberg would go into 'destroy mode' subjecting your business to the 'wrath of Mark,'" the attorneys general wrote, quoting from emails by competitors and investors. The chief executive was so afraid of losing out to rivals that he "sought to extinguish or impede, rather than outperform or out-innovate, any competitive threat." He spied on competitors, and he broke commitments to the founders of Instagram and WhatsApp soon after the start-ups were acquired, the states' complaint further alleged.

At Zuckerberg's side throughout was Sheryl Sandberg, the former Google executive who converted his technology into a profit powerhouse using an innovative and pernicious advertising business that was "surveilling" users for personal data. Facebook's ad business was predicated on a dangerous feedback loop: the more time users spent on the site, the more data Facebook mined. The lure was free access to the service, but consumers bore steep costs in other ways. "Users do not pay a cash price to use Facebook. Instead, users exchange their time, attention, and personal data for access to Facebook's services," the states' complaint asserted.

It was a growth-at-any-cost business strategy, and Sandberg was the industry's best at scaling the model. Intensely organized, analytical, hardworking, and with superior interpersonal skills, she was the perfect foil for Zuckerberg. She oversaw all the departments that didn't interest him—policy and communication, legal, human resources, and revenue creation. Drawing on years of public speaking training, and on political consultants to curate her public persona, she was the palatable face of Facebook to investors and the public, distracting attention from the core problem.

"It's about the business model," one government official said in an interview. Sandberg's behavioral advertising prototype treated human data as financial instruments bartered in markets like corn or pork belly futures. Her handiwork was "a contagion," the official added, echoing the words of academic and activist Shoshana Zuboff, who a year earlier had described Sandberg as playing "the role of Typhoid Mary, bringing surveillance capitalism from Google to Facebook, when she signed on as Mark Zuckerberg's number two."

With scant competition to force the leaders to consider the well-being of their customers, there was "a proliferation of misinformation and violent or otherwise objectionable content on Facebook's properties," the attorneys general alleged in their complaint. Even when faced with major impropriety such as Russia's disinformation campaign and the data privacy scandal involving Cambridge Analytica, users didn't leave the site because there were few alternatives, the regulators maintained. As James succinctly described, "Instead of competing on the merits, Facebook used its power to suppress competition so it could take advantage of users and make billions by converting personal data into a cash cow."

When the FTC and states came down with their landmark lawsuits against Facebook, we were nearing completion of our own investigation of the company, one based on fifteen years of reporting, which has afforded us a singular look at Facebook from the inside. Several versions of the Facebook story have been told in books and film. But despite being household names, Zuckerberg and Sandberg remain enigmas to the public, and for good reason. They are fiercely protective of the images they've cultivated—he, the technology visionary and philanthropist; she, business icon and feminist—and have surrounded the inner workings of "MPK," the shorthand employees use to describe the headquarters' campus in Menlo Park, with its moat of loyalists and culture of secrecy.

Many people regard Facebook as a company that lost its way: the classic Frankenstein story of a monster that broke free of its creator. We take a different point of view. From the moment Zuckerberg and Sandberg met at a Christmas party in December 2007, we believe, they sensed the potential to transform the company into the global power it is today. Through their partnership, they methodically built a business model that is unstoppable in its growth—with $85.9 billion in revenue in 2020 and a market value of $800 billion—and entirely deliberate in its design.

We have chosen to focus on a five-year period, from one U.S. election to another, during which both the company's failure to protect its users and its vulnerabilities as a powerful global platform were exposed. All the issues that laid the groundwork for what Facebook is today came to a head within this time frame.

It would be easy to dismiss the story of Facebook as that of an algorithm gone wrong. The truth is far more complex.

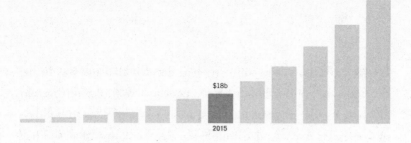

$18b

2015

Chapter 1

Don't Poke the Bear

It was late at night, hours after his colleagues at Menlo Park had left the office, when the Facebook engineer felt pulled back to his laptop. He had enjoyed a few beers. Part of the reason, he thought, that his resolve was crumbling. He knew that with just a few taps at his keyboard, he could access the Facebook profile of a woman he had gone on a date with a few days ago. The date had gone well, in his opinion, but she had stopped answering his messages twenty-four hours after they parted ways. All he wanted to do was peek at her Facebook page to satisfy his curiosity, to see if maybe she had gotten sick, gone on vacation, or lost her dog—anything that would explain why she was not interested in a second date.

By 10 p.m., he had made his decision. He logged on to his laptop and, using his access to Facebook's stream of data on all its users, searched for his date. He knew enough details—first and last name, place of birth, and university—that finding her took only a few minutes. Facebook's internal systems had a rich repository of information, including years of private conversations with friends over Facebook Messenger, events attended, photographs uploaded (including those she had deleted), and posts she had commented or clicked on. He saw the categories in which Facebook had placed her for advertisers: the company had decided that she was in her thirties, was politically left of center, and led an active lifestyle. She had a wide range of interests, from a love of dogs to holidays in Southeast Asia. And through the Facebook app that she had installed on her phone, he saw her real-time location. It was more information than the engineer could possibly have gotten over the course of a dozen dinners. Now, almost a week after their first date, he had access to it all.

Facebook's managers stressed to their employees that anyone discovered taking advantage of their access to data for personal means, to look up a friend's account or that of a family member, would be immediately fired. But the managers also knew there were no safeguards in place. The system had been designed to be open, transparent, and accessible to all employees. It was part of Zuckerberg's founding ethos to cut away the red tape that slowed down engineers and prevented them from producing fast, independent work. This rule had been put in place when Facebook had fewer than one hundred employees. Yet, years later, with thousands of engineers across the company, nobody had revisited the practice. There was nothing but the goodwill of the employees themselves to stop them from abusing their access to users' private information.

During a period spanning January 2014 to August 2015, the engineer who looked up his onetime date was just one of fifty-two Facebook employees fired for exploiting their access to user data. Men who looked up the Facebook profiles of women they were interested in made up the vast majority of engineers who abused their privileges. Most of the employees who took advantage of their access did little more than look up users' information. But a few took it much further. One engineer used the data to confront a woman who had traveled with him on a European vacation; the two had gotten into a fight during the trip, and the engineer tracked her to her new hotel after she left the room they had been sharing.

Another engineer accessed a woman's Facebook page before they had even gone on a first date. He saw that she regularly visited Dolores Park, in San Francisco, and he found her there one day, enjoying the sun with her friends.

The fired engineers had used work laptops to look up specific accounts, and this unusual activity had triggered Facebook's systems and alerted the engineers' managers to their transgressions. Those employees were the ones who were found out after the fact. It was unknown how many others had gone undetected.

The problem was brought to Mark Zuckerberg's attention for the first time in September 2015, three months after the arrival of Alex Stamos, Facebook's new chief security officer. Gathered in the CEO's conference room, "the Aquarium," Zuckerberg's top executives had braced themselves for potentially bad news: Stamos had a reputation for blunt speech and high standards. One of the first objectives he had set out when he was hired that summer was a comprehensive evaluation of Facebook's current state of security. It would be the first such assessment ever completed by an outsider.

Among themselves, the executives whispered that it was impossible to make a thorough assessment within such a short period of time and that whatever report Stamos delivered would surely flag superficial problems and give the new head of security some easy wins at the start of his tenure. Everyone's life would be easier if Stamos assumed the posture of boundless optimism that pervaded Facebook's top ranks. The company had never been doing better, with ads recently expanded on Instagram and a new milestone of a billion users logging on to the platform every day. All they had to do was sit back and let the machine continue to hum.

Instead, Stamos had come armed with a presentation that detailed problems across Facebook's core products, workforce, and company structure. The organization was devoting too much of its security efforts to protecting its website, while its apps, including Instagram and WhatsApp, were being largely ignored, he told the group. Facebook had not made headway on its promises to encrypt user data at its centers—unlike Yahoo, Stamos's previous employer, which had moved quickly to start securing the information in the two years since National Security Agency whistleblower Edward Snowden revealed that the government was likely spying on user data as it sat unprotected within the Silicon Valley companies. Facebook's security responsibilities were scattered across the company, and according to the report Stamos presented, the company was "not technically or culturally prepared to play against" its current level of adversary.

Worst of all, Stamos told them, was that despite firing dozens of employees over the last eighteen months for abusing their access, Facebook was doing nothing to solve or prevent what was clearly a systemic problem. In a chart, Stamos highlighted how nearly every month, engineers had exploited the tools designed to give them easy access to data for building new products to violate the privacy of Facebook users and infiltrate their lives. If the public knew about these transgressions, they would be outraged: for over a decade, thousands of Facebook's engineers had been freely accessing users' private data. The cases Stamos highlighted were only the ones the company knew about. Hundreds more may have slipped under the radar, he warned.

Zuckerberg was clearly taken aback by the figures Stamos presented, and upset that the issue had not been brought to his attention sooner. "Everybody in engineering management knew there were incidents where employees had inappropriately managed data. Nobody had pulled it into one place, and they were surprised at the volume of engineers who had abused data," Stamos recalled.

Why hadn't anyone thought to reassess the system that gave engineers access to user data? Zuckerberg asked. No one in the room pointed out that it was a system that he himself had designed and implemented. Over the years, his employees had suggested alternative ways of structuring data retention, to no avail. "At various times in Facebook's history there were paths we could have taken, decisions we could have made, which would have limited, or even cut back on, the user data we were collecting," said one longtime employee, who joined Facebook in 2008 and worked across various teams within the company. "But that was antithetical to Mark's DNA. Even before we took those options to him, we knew it wasn't a path he would choose."

Facebook's executives, including those in charge of the engineering ranks, like Jay Parikh and Pedro Canahuati, touted access as a selling point to new recruits on their engineering teams. Facebook was the world's biggest testing lab, with a quarter of the planet's population as its test subjects. The managers framed this access as part of Facebook's radical transparency and trust in its engineering ranks. Did a user enjoy the balloons on the prompt to wish her brother a happy birthday, or did an emoji of a birthday cake get a higher response rate? Instead of going through a lengthy and bureaucratic process to find out what was working, engineers could simply open up the hood and see for themselves, in real time. But Canahuati warned engineers that access to that data was a privilege. "We had no tolerance for the abuse, which is why the company had always fired every single person found to be improperly accessing data," he said.

Stamos told Zuckerberg and the other executives that it was not enough to fire employees after the fact. It was Facebook's responsibility, he argued, to ensure that such privacy violations never happened to begin with. He asked permission to change Facebook's current system to revoke private data access from the majority of engineers. If someone needed information on a private individual, they would have to make a formal request through the proper channels. Under the system then in place, 16,744 Facebook employees had access to users' private data. Stamos wanted to bring that number down to fewer than 5,000. For the most sensitive information, like GPS location and password, he wanted to limit access to under 100 people. "While everyone knew there was a large amount of data accessible to engineers, nobody had thought about how much the company had grown and how many people now had access to that data," Stamos explained. "People were not paying attention."

Parikh, Facebook's head of engineering, asked why the company had to upend its entire system. Surely, safeguards could be put in place that limited how much information an engineer accessed, or that sounded alarms when engineers appeared to be looking up certain types of data. The changes being suggested would severely slow down the work of many of the product teams.

Canahuati, director of product engineering, agreed. He told Stamos that requiring engineers to submit a written request every time they wanted access to data was untenable. "It would have dramatically slowed work across the company, even work on other safety and security efforts," Canahuati pointed out.

Changing the system was a top priority, Zuckerberg said. He asked Stamos and Canahuati to come up with a solution and to update the group on their progress within a year. But for the engineering teams, this would create serious upheaval. Many of the executives in the room grumbled privately that Stamos had just persuaded their boss to commit to a major structural overhaul by presenting a worst-case scenario.

One executive was noticeably absent from the September 2015 meeting. Only four months had passed since the death of Sheryl Sandberg's husband. Security was Sandberg's responsibility, and Stamos technically fell under her purview. But she had never suggested, nor been consulted about, the sweeping changes he was proposing.

Stamos prevailed that day, but he made several powerful enemies.

Late in the evening on December 8, 2015, Joel Kaplan was in the business center of a hotel in New Delhi when he received an urgent phone call from MPK. A colleague informed him that he was needed for an emergency meeting.

Hours earlier, Donald J. Trump's campaign had posted on Facebook a video of a speech the candidate had made in Mount Pleasant, South Carolina. In it, Trump promised to take a dramatically harder line against terrorists, and then he linked terrorism to immigration. President Obama, he said, had treated illegal immigrants better than wounded warriors. Trump would be different, the presidential candidate assured the crowd. "Donald J. Trump is calling for a total and complete shutdown of Muslims entering the United States until our country's representatives can figure out what the hell is going on," he announced. The audience exploded with cheers.

Trump had made inflammatory positions on race and immigration central to his presidential bid. His campaign's use of social media threw gas on the flames. On Facebook, the video of the anti-Muslim speech quickly generated more than 100,000 "likes" and was shared 14,000 times.

The video put the platform in a bind. It was unprepared for a candidate like Trump, who was generating a massive following but also dividing many of its users and employees. For guidance on this, Zuckerberg and Sandberg turned to their vice president of global public policy, who was in India trying to salvage Zuckerberg's free internet service program.

Kaplan dialed into a videoconference with Sandberg, Head of

Policy and Communications Elliot Schrage, Head of Global Policy Management Monika Bickert, and a few other policy and communications officials. Kaplan was thirteen and a half hours ahead of his colleagues at headquarters and had been traveling for days. He quietly watched the video and listened to the group's concerns. Zuckerberg, he was told, had made clear that he was concerned by Trump's post and thought there might be an argument for removing it from Facebook.

When Kaplan finally weighed in, he advised the executives against acting hastily. The decision on Trump's anti-Muslim rhetoric was complicated by politics. All those years of financial and public support for Democrats had dimmed Facebook's image among Republicans, who were growing distrustful of the platform's political neutrality. Kaplan was not part of Trump's world, but he saw Trump's campaign as a real threat. Trump's large following on Facebook and Twitter exposed a gaping divide within the Republican Party.

Removing the post of a presidential candidate was a monumental decision and would be seen as censorship by Trump and his supporters, Kaplan added. It would be interpreted as another sign of liberal favoritism toward Trump's chief rival, Hillary Clinton. "Don't poke the bear," he warned.

Sandberg and Schrage weren't as vocal on what to do with Trump's account. They trusted Kaplan's political instincts; they had no connections to Trump's circle and no experience with his brand of shock politics. But some officials on the conference line that day were aghast. Kaplan seemed to be putting politics above principle. He was so obsessed with steadying the ship that he could not see that Trump's comments were roiling the sea, as one person on the call described it.

Several senior executives spoke up to agree with Kaplan. They expressed concern about the headlines and the backlash they would face from shutting down comments made by a presidential candidate. Trump and his followers already viewed leaders like Sandberg and Zuckerberg as part of the liberal elite, the rich and powerful gatekeepers of information that could censor conservative voices with their secret algorithms. Facebook had to appear unbiased. This was essential to protecting its business.

The conversation turned to explaining the decision. The post could be seen as violating Facebook's community standards. Users had flagged the Trump campaign account for hate speech in the past, and multiple strikes were grounds for removing the account entirely. Schrage, Bickert, and Kaplan, all Harvard Law grads, labored to conjure legal arguments that would justify the decision to allow the post. They were splitting hairs on what constituted hate speech, right down to Trump's use of grammar.

"At one point, they joked that Facebook would need to come up with a version of how a Supreme Court Justice once defined pornography, 'I know it when I see it,'" recalled an employee involved in the conversation. "Was there a line they could draw in the sand for something Trump might say to get himself banned? It didn't seem wise to draw that line."

Facebook technically barred hate speech, but the company's definition of what constituted it was ever evolving. What it took action on differed within nations, in compliance with local laws. There were universal definitions for banned content on child pornography and on violent content. But hate speech was specific not just to countries but to cultures.

As the executives debated, they came to realize that they wouldn't have to defend Trump's language if they came up with a workaround. The group agreed that political speech could be protected under a "newsworthiness" standard. The idea was that political speech deserved extra protection because the public deserved to form their own opinions on candidates based on those candidates' unedited views. The Facebook executives were creating the basis for a new speech policy as a knee-jerk reaction to Donald Trump. "It was bullshit," one employee recalled. "They were making it up on the fly."

This was a critical moment for Joel Kaplan in terms of proving his value. Though unpopular to some on the call, he was providing crucial advice on a growing threat coming from Washington.

When Sandberg arrived at Facebook in 2008, the company had been neglecting conservatives. It was a critical oversight: where regulation over data collection was concerned, Republicans were Facebook's allies. When the House of Representatives flipped to a Republican majority in 2010, Sandberg hired Kaplan to balance the heavily Democratic ranks of the lobbying office and to change the perception in Washington that the company favored Democrats.

Kaplan came with sterling conservative credentials. A former deputy chief of staff to President George W. Bush, he was also a former U.S. Marine artillery officer and Harvard Law School graduate who had clerked for Supreme Court justice Antonin Scalia. He was the antithesis of the typical Silicon Valley liberal techie and, at forty-five, a couple of decades older than much of the staff at MPK. (He and Sandberg had met in 1987, during their freshman year at Harvard. They dated briefly and remained friends after their relationship ended.)

Kaplan was a workaholic who, like Sandberg, prized organization. At the White House, he had kept a trifold whiteboard in his office with lists of all the hot-button issues facing the administration: the auto bailout, immigration reform, and the financial crisis. His job was to manage complex policy issues and prevent problems from reaching the Oval Office. He occupied a similar role at Facebook. His mandate was to protect the business model from government interference, and to that end, he was an excellent employee.

In 2014, Sandberg had promoted Kaplan to lead global policy

in addition to Washington lobbying. For the past two years, Facebook had been preparing for a possible Republican administration after Obama. But Trump threw them off course. He was not of the Republican establishment. Kaplan's political capital seemed worthless when it came to the former reality TV star.

And while Trump was creating new headaches for Facebook, he was also a power user and important advertiser. From the start of Trump's presidential campaign, his son-in-law, Jared Kushner, and digital manager, Brad Parscale, put the majority of their media funds into the social network. They focused on Facebook because of its cheap and easy targeting features for amplifying campaign ads. Parscale used Facebook's microtargeting tools to reach voters by matching the campaign's own email lists with Facebook's user lists. He worked with Facebook employees who were embedded in Trump's New York City campaign headquarters to riff on Hillary Clinton's daily speeches and to target negative ads to specific audiences. They bought thousands of postcard-like ads and video messages. They were easily reaching bigger audiences than on television, and Facebook was an eager partner. Trump became an inescapable presence on the platform.

The 2016 U.S. presidential election would stamp out any doubts about the importance of social media in political campaigns. By early 2016, 44 percent of all Americans said they got their news about candidates from Facebook, Twitter, Instagram, and YouTube.

For nearly a decade, Facebook held an informal, company-wide meeting at the end of each week, known as "Questions and Answers," or Q&A. Its format was simple, and fairly standard in the industry: Zuckerberg would speak for a short time and then answer questions that had been voted on by employees from among those they'd submitted in the days ahead of the meeting. Once the questions that had received the most votes had been addressed, Zuckerberg would take unfiltered questions from the audience. It was more relaxed than Facebook's quarterly, company-wide meeting known as the "all-hands," which had a more rigid agenda and featured programs and presentations.

A couple hundred employees attended the meeting in Menlo Park, and thousands more watched a livestream of the meeting from Facebook's offices around the world. In the lead-up to the Q&A following Trump's Muslim ban speech, employees had been complaining in their internal Facebook groups—known as "Tribes"—that the platform should have removed Trump's speech from the site. In the broader forums where more professional discussions took place—known as "Workplace groups"—people asked for a history of how Facebook had treated government officials on the site. They were angry that Facebook's leaders hadn't taken a stand against what they viewed as clearly hate speech.

An employee stepped up to a microphone stand, and people grew quiet. Do you feel an obligation to take down the Trump campaign video calling for the ban on Muslims? he asked. The targeting of Muslims, the employee said, appeared to violate Facebook's rule against hate speech.

Zuckerberg was used to fielding hard questions at Q&As. He had been confronted about ill-conceived business deals, the lack of diversity in company staff, and his plans to conquer competition. But the employee in front of him posed a question on which his own top ranks could not find agreement. Zuckerberg fell back on one of his core talking points. It was a hard issue, he said. But he was a staunch believer in free expression. Removing the post would be too drastic.

It was a core libertarian refrain Zuckerberg would return to again and again: the all-important protection of free speech as laid out in the First Amendment of the Bill of Rights. His interpretation

was that speech should be unimpeded; Facebook would host a cacophony of sparring voices and ideas to help educate and inform
its users. But the protection of speech adopted in 1791 had been
designed specifically to promote a healthy democracy by ensur-
ing a plurality of ideas without government restraint. The First
Amendment was meant to protect society. And ad targeting that
prioritized clicks and salacious content and data mining of users
was antithetical to the ideals of a healthy society. The dangers present in Facebook's algorithms were "being co-opted and twisted by
politicians and pundits howling about censorship and miscasting
content moderation as the demise of free speech online," in the
words of Renée DiResta, a disinformation researcher at Stanford's
Internet Observatory. "There is no right to algorithmic amplification. In fact, that's the very problem that needs fixing."

It was a complicated issue, but to some, at least, the solution
was simple. In a blog post on the Workplace group open to all
employees, Monika Bickert explained that Trump's post wouldn't
be removed. People, she said, could judge the words for themselves.

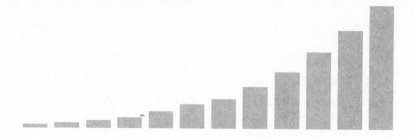

Chapter 2

The Next Big Thing

It's impossible to understand how Facebook arrived at its crisis point without looking back at how far the company had come, and how quickly.

The first time Mark Zuckerberg saw a website called "the Facebook," it had been conceived, coded, and named by someone else. It was a goodwill project meant to help friends connect with one another. It was free. And Zuckerberg's first instinct was to break it.

In September 2001, Zuckerberg was a seventeen-year-old senior at Phillips Exeter Academy, the prestigious boarding school in New Hampshire that had helped shape future government and industry leaders for more than two centuries. The son of a dentist, Zuckerberg had a different pedigree than many of his peers, who were descendants of former heads of state and corporate chieftains. But the gangly teenager quickly found his place, thriving in the school's Latin program and Comp Sci classes and establishing himself as the campus computer geek. Fueled by Red Bull and Cheetos, he led other students on all-night coding binges, trying to hack into the school's systems or creating algorithms to speed up assignments. Sometimes Zuckerberg set up programming races; usually, he won.

At the time, the Student Body Council was planning to put the school's student directory online. "The Photo Address Book," a laminated paperback that listed students' names, phone numbers, addresses, and head shots, was an Exeter institution. "The Facebook," as everyone called it, had hardly changed for decades.

The initiative had come from council member Kristopher Tillery, who was in the same year as Zuckerberg. As a coder, Tillery considered himself a dabbler, but he was fascinated by companies like Napster and Yahoo, which had become widely popular among his fellow students. He wanted to make Exeter, a school that dated to 1781, feel cool and modern. What better way, he figured, than to upload the Facebook directory to the internet?

He never expected it to take off the way it did. The ease of bringing up the profile of any classmate with just the tap of a few keys was novel. It raised the art of pranking to a new level. Anchovy pizzas were sent to dorm rooms. Students pretending to be school officials would call up classmates to alert them to a flood in the building or to accuse them of plagiarizing a paper.

But before long, students started complaining to Tillery about a problem: the page for Mark Zuckerberg wasn't working. Whenever students tried to open Zuckerberg's entry on the site, their browsers crashed. The window they were using closed, and sometimes their computers froze and had to be restarted.

When Tillery investigated, he found that Zuckerberg had inserted a line of code into his own profile that was causing the crashes. It was easily fixed. *Of course it was Mark*, Tillery thought. "He was very competitive, and very, very, very smart. He wanted to see if he could push what I was doing a little further. I saw it as a test and just him flagging for people that his skills were, well, better than mine."

The Facebook origin story—how Zuckerberg got drunk one night at Harvard two years later and started a blog to rate his female classmates—is well worn at this point. But what often gets omitted from the mythology is the fact that while many students immediately embraced Zuckerberg's creation, called "FaceMash," others were alarmed by the invasion of privacy. Just days after FaceMash was launched, two student groups at Harvard, Fuerza Latina, a pan-Latino cultural group, and the Association of Black Harvard Women, emailed Zuckerberg to voice concerns about his site.

Zuckerberg responded directly to both groups, explaining that the popularity of the site had come as a surprise. "I understood that some parts were still a little sketchy and I wanted some more time to think about whether or not this was really appropriate to release to the Harvard community," he wrote in an email he knew would be made public. He added: "This is not how I meant for things to go, and I apologize for any harm done as a result of my neglect to consider how quickly the site would spread and its consequences thereafter."

Harvard's computer services department filed a complaint, alleging that Zuckerberg had violated copyrights; he had also potentially violated guidelines around student IDs. When it came time for his hearing, Zuckerberg repeated the explanation that he'd given the student groups. The site had been a coding experiment. He was interested in the algorithms and in the computer science behind what made the site work. He never expected the project to go viral, he insisted, and he apologized if any of his fellow students felt that their privacy had been violated. In what would become a familiar pattern, he got off with a scolding and agreed to meet regularly with a university counselor.

Then he returned to the business of creating a private, student-only social network. Several of his peers were circling the same idea, most notably Cameron and Tyler Winklevoss, who along with the equally well-connected Divya Narendra had approached Zuckerberg about writing code for their effort. But Zuckerberg was focused on one student already well ahead of him. Earlier that fall, a junior named Aaron Greenspan introduced a networking site called "the Face Book." It was a simple site, designed to look professional. Greenspan's idea was to create a resource that could be useful to professors or job hunters. But the earliest iterations of "the Face Book" drew criticism for allowing students to post personal details about classmates, and the *Harvard Crimson* slammed the project as a possible security risk. The site quickly stalled as a result of the pushback.

Greenspan reached out to Zuckerberg after hearing his name around campus, and the two developed a competitive friendship. When an instant message popped up from Zuckerberg on January 8, 2004, Greenspan was surprised; he hadn't given Zuckerberg his AOL username. The two had sat through an awkward dinner in Kirkland House earlier that night, during which Zuckerberg had fended off Greenspan's questions on the type of projects he was interested in pursuing next. But over chats, Zuckerberg floated the idea of combining his in-development social network with Greenspan's project. Greenspan pushed back on the suggestion that he redesign his site and asked Zuckerberg if he wanted to incorporate what he was building into what Greenspan had already launched.

"It would be sort of like how delta has song airlines," Greenspan wrote.

"Delta owns song airlines," Zuckerberg responded.

Zuckerberg wasn't enthusiastic about tailoring his ambition to fit what Greenspan had built, and he wondered aloud if they might become competitors instead. Zuckerberg wanted his creation to be less formal. Users were more likely to talk about hobbies or their favorite music in their living rooms than in their offices. If the social network felt "too functional," he told Greenspan, users wouldn't share as much. He wanted to design a place to "waste time."

He also revealed that he was already thinking about the ways personal data might be repurposed. Greenspan's site asked users to share specific bits of information toward a specific purpose. Phone numbers allowed classmates to connect; addresses provided meeting places for study groups. "In a site where people give personal information for one thing, it then takes a lot of work and precaution to use that information for something else," Zuckerberg wrote. He wanted users to share data in an open-ended way, expanding and diversifying the kinds of information he could collect.

The two discussed sharing a common database of users, along with the idea of automatically registering students for both versions of Thefacebook (as the name was now being styled) when they signed up for one. Their conversations ebbed and flowed, but Zuckerberg ultimately decided that his own project offered unique features, and he preferred his more casual design.

Zuckerberg intuited that the success of his site depended on the willingness of his fellow students to share intimate details about themselves. He was fascinated by human behavior; his mother was a practicing psychiatrist before she had children, and he was a psychology major. He focused on how easily students shared personal information. Every drunken photo, every pithy joke, and every quotable story was free content. That content would drive more people to join Thefacebook to see what they were missing. The challenge was to make the site a place where users mindlessly scrolled: "I kind of want to be the new MTV," he told friends. The more time that users spent on Thefacebook, the more they would reveal about themselves, intentionally or not. The friends whose pages they visited, the frequency with which they visited those pages, the hookups they admitted to—every connection accelerated Zuckerberg's vision of an expansive web of social interactions.

"Mark was acquiring data for the sake of data because, I think he is a lot like me. I think he saw that the more data you had, the more accurately you could build a model of the world and understand it," said Greenspan, who continued to keep in touch with Zuckerberg after he launched his competing site. "Data is extremely powerful, and Mark saw that. What Mark ultimately wanted was power."

Zuckerberg's site assured students that because the network was limited to Harvard, it was private by design. But Facebook's earliest terms of service made no mention of how users' personal details (which they hadn't yet come to think of as their individual data) could be used. In later years, Zuckerberg would tout, over and over again, the power of his invention to connect people—the whole world, in fact. But in those early days, his focus was altogether different. In one online chat, he made clear just how much access he had to the data he had accumulated. Zuckerberg began the conversation with a boast, telling one friend that if he ever needed information on anyone at Harvard, he should just say the word:

ZUCK: i have over 4000 emails, pictures, addresses, sns
FRIEND: what!? how'd you manage that one?
ZUCK: people just submitted it
ZUCK: i don't know why
ZUCK: they "trust me"
ZUCK: dumb fucks

In January 2005, Zuckerberg shuffled into a small conference room at the *Washington Post* for a business meeting with the chairman of one of the oldest and most venerated newspapers in America. Zuckerberg was about to celebrate the one-year anniversary of his social media company, Thefacebook. More than a million people were using his site, putting the twenty-year-old in rarefied company. Zuckerberg accepted his celebrity status among like-minded techies, but going into this particular meeting, he was visibly nervous.

He was uncomfortable in the political corridors of Washington, DC, and unfamiliar with the clubby world of East Coast media. Just six months earlier, he had moved to Palo Alto, California, with a few friends from Harvard. What had begun as a summer vacation experiment—running Thefacebook from a five-bedroom ranch house with a zipline strung over the backyard pool—had turned into an extended leave of absence from school, one spent meeting venture capitalists and entrepreneurs who were running some of the most exciting tech companies in the world.

"He was like a very nerdy movie star," observed one friend who worked for the start-up and frequented the Palo Alto home Zuckerberg and his roommates had dubbed "Casa Facebook." "Facebook was still small, by Silicon Valley standards, but a lot of people already saw him as the next big thing."

The ideas that Zuckerberg would have been absorbing during his junior year at Harvard were replaced with the philosophies of entrepreneurs like Peter Thiel, the cofounder of PayPal, who had invested $500,000 in Thefacebook in August 2004, and Marc Andreessen, Netscape's cofounder. As two of the most powerful men in Silicon Valley, they did more than just create and invest in new start-ups: they shaped the ethos of what it meant to be a tech ingénue. That ideology was rooted in a version of libertarianism that embraced innovation and free markets and scorned the overreach of government and regulations. At the core was a belief in individual autonomy, inspired by philosophers and writers like John Stuart Mill and Ayn Rand and their advocacy of rationalism and skepticism toward authority. The driving goals were progress and profits. The businesses emerging from Silicon Valley were redefining the old ways of doing things, the inefficient and bad habits that needed breaking. (In 2011, Thiel would endow a fellowship to pay college students to drop out to take on apprenticeships and start companies.)

The education was informal. "I never saw Mark reading a book, or expressing any interest in books," said a friend, who recalled many late-night video game marathons in which vague ideas around warfare and battles were used as allegories for business. "He was absorbing ideas that were floating around at the time, but he didn't have a keen interest in the origin of those ideas. And he definitely didn't have a broader interest in philosophy, political thought, or economics. If you asked, he would say he was too busy taking over the world to read."

Zuckerberg had few contacts outside the world of tech enthusiasts and engineers. But during the holidays, his classmate Olivia Ma had convinced her father, a vice president at the *Washington Post* in charge of new ventures, to meet the precocious young coder whose website was sweeping college campuses across the country. Her father, impressed, set up a meeting at the paper's DC headquarters.

Zuckerberg showed up at the paper's offices wearing jeans and a sweater, accompanied by Sean Parker, the creator of Napster, who had become Facebook's new president months earlier. Last to arrive to the small conference room where they had been escorted was Donald Graham, chairman of the *Post* and the third-generation leader of the family newspaper.

Graham was a fixture in the society pages of New York City and Washington, having grown up mingling with the families of John F. Kennedy and Lyndon B. Johnson and business tycoons like Warren Buffett. Under his leadership, the *Post* had won more than twenty Pulitzer Prizes and other journalism awards, building on the reputation it had earned for its groundbreaking Watergate coverage. But Graham could see the looming threat of digital media. Advertisers were excited about the explosion of growth in internet use, and sites like Google and Yahoo were scraping stories from CNN, the *Post*, and other media to draw audiences to their platforms and away from the news organizations' own nascent websites.

Graham wanted to reach a new generation of readers. Unlike many of his counterparts in the music industry and Hollywood, he hadn't taken a hostile stance toward the tech platforms; instead, he was seeking information and potential partnerships. He had already talked to Jeff Bezos about Amazon's distribution of books, and now he was curious about this young techie on a leave of absence from his alma mater. "I was not somebody who understood technology deeply, but I wanted to learn," Graham recalled.

The kid struck him as extremely awkward and shy. Zuckerberg didn't seem to blink as he haltingly explained to Graham, almost forty years his senior, how Thefacebook worked. Students at Harvard created their own pages with basic information: their name, class, dormitory, college clubs, hometown, majors. One student could look up another and ask them to be a "friend." Once connected, the two could comment on each other's pages and post messages. "Who is going to Widener Library on Thursday night? Want to study for the Chem exam?" The site had a few ads from local Cambridge businesses, just enough to cover the bills for more tech equipment.

"Well, there goes the *Crimson*," Graham said, referring to Harvard's student newspaper. "Every pizza parlor in Cambridge is going to stop advertising in the *Crimson* because of you." With so many eyeballs gravitating to the social network and the relatively inexpensive cost of its ads, any travel, sports equipment, or computer company trying to reach college students would be foolish if they didn't advertise on the site, he added.

Zuckerberg laughed. Yes, he said. But it wasn't revenue he was really after, he explained—Graham would later note that Zuckerberg didn't seem to know the difference between profit and revenue—it was people. He wanted more users. He told Graham he had to race to expand to every college in the nation before some-one else did.

The platform was focused on pursuing scale and engagement. And unlike the *Post*, Zuckerberg had a long runway to build that audience without the pressure of making money. Graham was struck by the company's potential reach. He saw it not as a threat to the traditional newspaper business, but as a technology that could be a good partner as the *Post* navigated its future online; he had witnessed the trouble that the music and entertainment businesses had run into during the shift to digital. Twenty minutes into the conversation, he told Zuckerberg that Thefacebook was one of the best ideas he had heard in years. Within days of the meeting, he offered $6 million for a 10 percent stake in the company.

Parker liked the idea of an investment from a media firm. He felt he had been burned by venture capital investors while running Napster, and he didn't trust them. In contrast with Zuckerberg, Parker came across as a slick salesman; his preference for a media firm's involvement seemed somewhat ironic, given his history as the cofounder of the peer-to-peer music sharing service that was the subject of multiple piracy lawsuits by record labels. Neverthe-less, the three quickly hammered out the broad contours of a deal. There was no term sheet, just an oral agreement.

For the next several weeks, the lawyers at the *Washington Post* and Thefacebook went back and forth on an official contract. At one point, Thefacebook asked for more money and for Graham to sit on the board. In March, Zuckerberg called Graham. He had a "moral dilemma," he confessed. He had received an offer from the venture capital firm Accel Partners that more than doubled what Graham had put on the table.

Accel offered deep coffers without any of the fussiness, or Old Guard values, of traditional financing. There was no interest in pushing young founders like Zuckerberg on issues like profitability or accountability. Start-ups were encouraged to operate on losses of millions of dollars a month, as long as they were attracting customers and innovating. The strategy was simple: be first to market, grow like crazy, and figure out the money later.

Graham appreciated that Zuckerberg had been open about his dilemma. He told the young man to take the offer that made the most sense for his business. The advice chimed with Zuckerberg's competitive instincts. "When you met Mark, that was the first thing you took away about him," said the longtime friend who had frequented Casa Facebook. "He wasn't someone who liked to lose."

By the winter of 2005, Thefacebook had become one of the most talked-about companies in Silicon Valley. It wasn't just the number of users the platform seemed to be acquiring daily; it was also its gold mine of user data. Users were volunteering personal information from the moment they signed up for an account: hometowns, phone numbers, the schools they attended and the jobs they had, their favorite music and books. No other tech company was collecting comparable data of such depth and breadth. By the end of 2004, one million college students had joined the platform. Even more impressive, they were logging in to Thefacebook more than four times a day, and on every campus Zuckerberg approved for the site, the majority of students signed up.

Investors saw Zuckerberg as the next founder genius, one following in the footsteps of Gates, Jobs, and Bezos. "It was an idea that just permeated the Valley at the time, and it was this sense that you could not question the founder, that they were king," noted Kara Swisher, a journalist who watched the rise of Zuckerberg and closely followed the men who mentored him. "My impression of Zuckerberg was that he was an intellectual lightweight, and he was very easily swayed by Andreessen or Peter Thiel; he wanted to be seen as smart by them so he adopted the changes they suggested and the libertarian mindset they projected." It didn't hurt, Swisher added, that Zuckerberg had a restless drive that compelled him to do whatever it took to ensure his company would be successful.

He had already shown that killer instinct in locking down control of Thefacebook in the first year of its operation. At Harvard, Zuckerberg had generously handed out titles and recruited his friends for his new company; one of his classmates, Eduardo Saverin, assumed the title of cofounder in return for the modest sum he had invested in helping Thefacebook get off the ground. In July 2004, however, Zuckerberg incorporated a new company, which essentially purchased the LLC he had formed with Saverin. The arrangement allowed Zuckerberg to redistribute shares in a way that guaranteed him the majority, while sinking Saverin's stake from roughly 30 percent to less than 10 percent. Saverin protested the decision and later sued for compensation. The board at the time, which comprised two early investors, Jim Breyer of Accel and Peter Thiel, essentially served in an advisory capacity and afforded Zuckerberg wide latitude on such decisions.

In September 2005, the platform, which had jettisoned the definite article from its name to become just "Facebook," began adding high school students as well as continuing to expand to more universities. From its tiny headquarters just above a Chinese restaurant on Palo Alto's central drag, Zuckerberg rallied his employees to work ever-longer shifts to keep up with the demand. As the platform marked more than 5.5 million users at the end of the year, he began ending weekly meetings by pumping his fist in the air and yelling, "Domination!"

The investment and buyout offers kept coming in. Viacom, Myspace, Friendster—all made plays in the wake of the *Washington Post* and Accel proposals. Yahoo, which in June 2006 offered $1 billion, was the hardest to reject. Few tech start-ups as small as

Facebook, and with no profits, had been offered that type of figure before. A number of employees implored Zuckerberg to take the deal. His board and other advisers told him he could walk away from Facebook with half the sum—and the stature to do anything he wanted.

The Yahoo buyout offer forced Zuckerberg to think about his long-term vision. By July, he told Thiel and Breyer that he didn't know what he'd do with the money and that he'd probably just build another version of Facebook. And in truth, he'd realized the platform could be exponentially bigger. "It was the first point where we really had to look at the future," he said. He and cofounder Dustin Moskovitz decided that they could "actually go and connect more than just the ten million people who are in schools."

Zuckerberg's whole management team left in protest when he rejected the Yahoo deal. It was his lowest point as a business leader, he later recalled. It also marked a crossroads for the company. "The part that was painful wasn't turning down the offer. It was the fact that after that, huge amounts of the company quit because they didn't believe in what we were doing."

And yet the move bolstered Zuckerberg's reputation. His audacity brought new confidence in the company. He began poaching employees from Microsoft, Yahoo, and Google. "People wanted to work at Facebook. It had this aura that it was going to be a big thing," observed an employee who was among the first fifty hires. "You knew that if you had Facebook on your résumé, it was going to look good."

In spite of the company's rising profile, the culture remained scrappy. Everyone crowded around the same small desks, which were often littered with the coffee cups and chocolate wrappers of the people who had worked the previous shifts. Meetings were abruptly called off if engineers failed to show finished prototypes of their ideas, despite having no managers to weigh in or offer guidance. All-night hackathons in which employees, fueled by beer and energy drinks, coded new features were common.

Zuckerberg relished the coding sessions, but he spent most of his time working on an idea that he believed would propel Facebook ahead of its competitors, one that gave him the confidence to turn down Yahoo: a customized central landing page. Up until this point, if users wanted to view updates from their friends, they had to click through to each user's individual page. Facebook was like a straightforward directory then, with no connection between profiles or opportunities to communicate easily. The new feature, called "News Feed," would draw from posts, photos, and status updates that users had already entered into their Facebook profiles and reorganize them into a unified feed—essentially, a continually refreshing stream of information.

Despite the name, Zuckerberg envisioned News Feed as blurring the definition of traditional news. While a newspaper's editors determined the hierarchy of articles that ran on the front of a newspaper or a website's home page, Zuckerberg imagined a personalized hierarchy of "interesting-ness" that would dictate what each user saw in their individual version of the feed. First and foremost, users would want to see content about themselves, so any post, photo, or mention of a user should appear at the top of the user's News Feed. Next would be content about friends, in descending order from those with whom the user interacted the most. Content from joined pages and groups would follow.

As simple as News Feed looked in his notebooks, Zuckerberg knew it would be challenging to create. He needed someone who could help him develop an algorithm that could rank what users wanted to see. He turned to Ruchi Sanghvi, one of his earliest employees and engineers, to anchor the technical work. Overseeing the project was a group of managers Zuckerberg had recently hired, most notably Chris Cox.

Cox had been plucked from a graduate program at Stanford studying natural language processing, a field of linguistics that looked at how artificial intelligence could help computers process and analyze the way people spoke. With his buzz cut and perpetual tan, he looked like a surfer, but he sounded like a technologist. Cox was known as a top student at Stanford; his departure for a small start-up that was competing against the much larger and better-funded Myspace and Friendster confounded his professors and classmates. He had accepted the Facebook job without meeting Zuckerberg, but from the moment the two were introduced, there was a natural fit. Cox had charisma and a talent for putting his boss at ease, and he seemed to know intuitively what Zuckerberg would think about a product or design.

Cox was ideally positioned to translate Zuckerberg's vision for News Feed for other employees: Zuckerberg wanted people to stay connected to their friends by spending hours every day scrolling through News Feed. He also wanted them to stay connected to the site; the goal was to keep users logging as much active time on the platform as possible—a metric that would come to be known as "sessions." From an engineering perspective, the News Feed system was by far the most intense and complicated design Facebook had tackled. It took nearly a year to code, but its impact was immeasurable: News Feed would not only change the course of the platform's history but go on to inspire scores of other tech companies around the world to reimagine what people wanted to see on the internet.

Just after 1 a.m., Pacific Standard Time, on September 5, 2006, employees crowded into one corner of the office to watch News Feed go live. Outside of Facebook employees, and a handful of investors whom Zuckerberg had briefed, no one knew about Facebook's plans for a major overhaul. Some advisers had pleaded with Zuckerberg to do a soft launch. Zuckerberg ignored them.

Instead, when the designated hour came, people logging into Facebook from across the United States were suddenly shown a prompt informing them that the site was introducing a new feature. There was only one button for them to click, and it read "Awesome." Once the button was clicked, the old Facebook disappeared forever. Few users bothered to read the accompanying blog post from Sanghvi, which introduced the new features in a cheerful tone. Instead, they dived headfirst into Zuckerberg's creation. At least one user was not impressed: "News Feed sucks," read an early post. Zuckerberg and his engineers laughed it off. It would take people some time to get used to the new design, they thought. They decided to call it a night and go home to sleep.

But the morning brought angry users outside Facebook's offices on Emerson Street, and virtual protests to a Facebook group called "Students Against Facebook News Feed." The group was furious that relationship updates were suddenly being posted in what felt like a public message board. Why did Facebook need to broadcast that a relationship had gone from "Just friends" to "It's complicated"? they asked. Others were dismayed to see their summer vacation photos shared with the world. Though the feature built on information they had made public on the site, users were just now coming face-to-face with everything Facebook knew about them. The encounter was jarring.

Within forty-eight hours, 7 percent of Facebook users had joined the anti–News Feed group, which was created by a junior at Northwestern University. The company's investors panicked, with several calling Zuckerberg to ask him to turn off the new feature. The ensuing PR fallout seemed to support the suggestion: privacy advocates rallied against Facebook, decrying the new design as invasive. Protesters demonstrated outside the Palo Alto office and Zuckerberg was forced to hire Facebook's first security guard.

And yet, Zuckerberg found comfort in the numbers. Facebook's data told him that he was right: users were spending more time on the site than ever before. In fact, the Facebook group Students Against Facebook News Feed proved that the News Feed was a hit—users were joining the group because they were seeing it at the top of their News Feed. The more users who joined, the more Facebook's algorithms pushed it to the top of the feed. It was Facebook's first experience with the power of News Feed to insert something into the mainstream and create a viral experience for its users.

"When we watched people use it, people were really, really, really using it a lot," Cox recalled. "There was a ton of engagement, and it was growing." The experience confirmed Cox's dismissal of the initial public response as the kind of knee-jerk reaction that had accompanied the introduction of all new technologies throughout history. "When you go back and you look at the first radio, or the first time we talked about the telephone and everybody said this is going to invade our privacy to put telephone lines in our houses because now people will call and they'll know when I'm not home and they'll go break into my house," he said. "That's probably happened a few times, but on balance, telephones are probably good."

Still, Zuckerberg knew that he had to do something to calm the backlash against the platform. At the end of a day spent fielding calls from friends and investors, he decided to say he was sorry. Just before 11 p.m. on September 5, almost twenty-four hours after News Feed launched, the CEO posted an apology on Facebook titled, "Calm Down. Breathe. We Hear You." The 348-word note set the tone for how Zuckerberg would deal with crises going forward. "We are listening to all your suggestions about how to improve the product; it's brand new and still evolving," he wrote, before noting that nothing about users' privacy settings had changed. (Whether that was in fact the case, within weeks, Facebook's engineers would introduce tools allowing users to restrict access to some information.) Facebook wasn't forcing users to share anything they didn't want to share. If they weren't happy with what they'd posted, well . . . they shouldn't have posted it in the first place. Ultimately, the note read less like an apology than an admonition from an exasperated parent: This food is good for you. Someday you'll thank me.

The *New York Times* had published its paper for more than one hundred years under the motto "All the News That's Fit to Print." Facebook was publishing its news under a different kind of motto: All the news from your friends that you never knew you wanted.

Almost immediately, the company ran into the issue of the lack of an editor, or defining principles. Newspapers drew on years of editorial judgment and institutional knowledge to determine what they would publish. The task of deciding what Facebook would and would not allow on its platform fell to a group of employees who had loosely assumed roles of content moderators, and they sketched out early ideas that essentially boiled down to "If something makes you feel bad in your gut, take it down." These guidelines were passed along in emails or in shared bits of advice in the office cafeteria. There were lists of previous examples of items Facebook had removed, but without any explanation or context behind those decisions. It was, at best, ad hoc.

This problem extended to advertising. The ads themselves were uninspiring—postage stamp boxes and banners across the site—and the small team that oversaw them generally accepted most submissions. When Director of Monetization Tim Kendall was hired, just before the launch of the News Feed, there were no set guidelines dictating acceptable ad content. And there wasn't a vetting process in place for Kendall and the ad team that reported to him. They were essentially making it up as they went along. "All policy decisions on content were totally organic and done as a response to problems," a former employee said.

In the summer of 2006, Kendall encountered his first difficult call. Middle East political groups were attempting to buy advertisements targeting college students that intended to stir up animosities on a given side of the Palestinian-Israeli conflict. The ads were graphic, featuring gruesome images, including photos of dead children. The team agreed that they didn't want to run the ads, but they were unsure how to justify their refusal to the political groups. Kendall quickly crafted a few lines stating that the company would not accept advertisements that incited hate or violence. He did not get approval from his bosses, and Zuckerberg did not weigh in.

It was all very informal; there was no legal vetting of the one-sheet policy paper. Content policy was far down on the company's list of priorities. "We didn't understand what we had in our hands at the time. We were one hundred people at the company with five million users, and speech was not on our radar," another employee recalled. "Mark was focused on growth, experience, and product. We had the attitude that, how could this frivolous college website have any serious issues to grapple with?"

The CEO's attention was trained elsewhere. Across the United States, high-speed internet access was growing. Better bandwidth at home and 24/7 internet access created fertile conditions for innovations in Silicon Valley—among which were new social networks focused on offering constant, ever-changing streams of information. Twitter, which had launched in July 2006, was on its way to reach a million users. It was used by everyone, ranging from the teenager next door to the man who would become the leader of the free world.

To stay ahead, Facebook would have to get bigger. Much bigger. Zuckerberg had set his sights on connecting every internet user in the world. But to get even close to that goal, he faced a daunting problem: how to monetize. In its fourth year, Facebook needed to come up with a plan to turn eyeballs into additional dollars. It needed to up its ad game. Investors were willing to tolerate rising costs and red ink—but only to a point. Expenses piled up as Zuckerberg hired more staff, rented more office space, and invested in servers and other equipment to keep pace with growth.

As Don Graham had sensed, Zuckerberg was a businessman with little enthusiasm for business. The year after Facebook was created, he reflected on his early experiences as a CEO in a documentary about Millennials by Ray Hafner and Derek Franzese. He mused aloud about the idea of finding someone else to handle all the parts of running a company that he found tiresome.

"The role of CEO changes with the size of the company," the then-twenty-one-year-old Zuckerberg says in the film. He is sitting on a couch in the Palo Alto office, barefoot and wearing a T-shirt and gym shorts. "In a really small start-up, the CEO is very often like the founder or idea guy. If you are a big company, the CEO is really just managing and maybe doing some strategy but not necessarily being the guy with the big ideas.

"So, like, it's a transition moving toward that and also, do I want to continue doing that? Or do I want to hire someone to do that and just focus more on like cool ideas? That's more fun," he says, chuckling.

$777m

2009

Chapter 3

What Business Are We In?

In the fall of 1989, more than a decade before Mark Zuckerberg arrived in Cambridge, Harvard professor Lawrence Summers walked into his Public Sector Economics course and scanned the room. It was a challenging class that delved into labor markets and the impact of social security and health care on broader economic markets. The thirty-four-year-old professor, a rising star in economics, could easily peg the suck-ups among the one hundred students in the lecture hall. They were the ones who sat at the front, trigger-ready to raise their hands. Summers found them annoying. But they were also typically the top performers.

So, months later, Summers was surprised when the top grade in the midterm exam went to a student who had barely caught his attention. He struggled to place the name. "Sheryl Sandberg? Who's that?" he asked a teaching assistant.

Oh, the junior with the mop of dark, curly hair and the oversize sweatshirts, he recalled. She sat at the far right of the room among a group of friends. She didn't speak up much, but she appeared to take copious notes. "She wasn't one of those kids desperate to get called on, raising both hands," he recalled.

As was his tradition, Summers invited the highest-scoring students to lunch at the Harvard Faculty Club, located at the edge of Harvard Yard, the historic heart of campus. Sandberg shone during the lunch, prepared with good questions. Summers liked how she had done her research. But he was also struck by how different she was from most of the overachievers in his classes.

"There are a lot of undergrads who run around Harvard whom I called 'closet presidents,'" Summers said. "They think they will be president of the U.S. someday and that they are really hot shit. Sheryl didn't hold herself or treat herself as hugely important."

At the time, Summers was beginning to draw national attention for his expertise on the intersection of government and economics; he had received his doctorate in economics at Harvard at the age of twenty-seven and had become one of the youngest tenured professors in the university's recent history. He was not easily impressed, but Sandberg stood out from her peers. She asked him to support Women in Economics and Government, a student-run group she had cofounded with female classmates. The club's goal was to increase the number of female faculty and students in the concentration; Sandberg rattled off data on the percentage of female economics majors, which was in the single digits. Summers accepted the invitation. She had clearly put a lot of thought into the club, and he appreciated her good manners and respectful demeanor.

"There are students who call me Larry before I tell them it's okay to call me Larry. Then there are students who I've said to call me Larry and I have to remind to call me Larry and not Professor Summers," he said. "Sheryl was the latter."

Summers left during Sandberg's senior year to become the chief economist for the World Bank, but he recruited her to Washington, DC, shortly after she received her diploma in the spring of 1991, to join him as a research assistant. (He had continued as her senior thesis adviser, but felt a "little guilty" for the limited time he could offer her from Washington.)

When Sandberg told her parents she was planning to return to her alma mater in the fall of 1993 to attend business school, they were skeptical. They had always assumed she'd end up in the public sector or at a nonprofit, and until recently, so had she. Her family was full of physicians and nonprofit directors. Her father was an ophthalmologist and, with her mother, had led the South Florida chapter of the Soviet Jewry movement, a national effort to raise awareness of the persecution of Jews in the Soviet Union. Sheryl's younger siblings followed their father's professional path; her brother became a renowned pediatric neurosurgeon, and her sister, a pediatrician. The World Bank made sense to them. Law school seemed like a more logical next step.

A senior colleague at the World Bank, Lant Pritchett, who dubbed Sandberg "a Mozart of human relations," had encouraged her to go to business school instead. "You should be leading people who go to law school," he said. Years later, Sandberg's father told Pritchett he was right.

Sandberg had a couple false starts in the ensuing years. In June 1994, in the middle of her MBA program, she married Brian Kraff, a Columbia Business School graduate and start-up founder in Washington, DC. The marriage ended after one year; Sandberg later told friends that she got married too young and that it wasn't a good match: Kraff seemed content to settle into a comfortable life in Washington, while she was more ambitious.

MBA in hand, she went to Los Angeles to work at the consulting firm McKinsey and Company, but she found she wasn't interested in management consulting. After a year, in late 1996, she rejoined Summers in Washington, at the Treasury Department. There, she encountered a stroke of good fortune. A constant stream of CEOs passed through the doors of Treasury to meet with Summers, and one of them was Eric Schmidt, a tech executive who was poised to take the reins at Google.

Sandberg was drawn to the dynamism of Silicon Valley. Companies like eBay were enabling millions of people to transform themselves into entrepreneurs. Yahoo and Google promised to expand knowledge across the world, bringing search and free communications tools like email and chat to Asia, Europe, and Latin America. Founded as a garage start-up in Seattle in 1995, Amazon was roiling the U.S. retail sector, with $2.7 billion in revenue in 2000.

In particular, Sandberg was curious about the profound impact tech was having on people's lives and the potential of companies to lead social movements. "Sheryl wants to have impact with a capital *I*," a friend explained. "Yes, money is important, but more important is the idea of doing something big and with impact."

Sandberg had traveled with Summers to meet with Schmidt, then the chief executive of the business software giant Novell. Unlike Wall Street and Fortune 500 executives, who used chauffeured Lincoln Town Cars and dressed in expensive custom-tailored suits, Schmidt picked them up from the San Francisco airport in his own car, wearing jeans. They ate at a local pizza place with Yahoo's founder, Jerry Yang, who had just become a billionaire. Sandberg was used to the ways of conventional business and political power, but here were these men with no interest in stuffy protocols exchanging both the ideas in their heads and the food on their plates with all the casualness in the world.

The memory stayed with her. In 2001, when Schmidt became the chief executive of Google, he offered her a job as a business unit manager. It was a vague title, and there were no business units at the time. Sandberg told Schmidt she was reluctant to join without a clear mandate. "Sheryl, don't be an idiot," Schmidt said. "If you're offered a seat on a rocket ship, get on, don't ask what seat." The company was the most exciting start-up at the time, its name already used as a verb for finding anything online, and it was destined to have one of the biggest IPOs on the market. But Sandberg was primarily taken with the vision of its founders, two former doctoral students at Stanford whose aim was to make the world's information universally accessible and useful. After Larry Page and Sergey Brin founded the search engine in 1998, they'd adopted an unofficial motto: "Don't Be Evil."

This sense of idealism resonated with Sandberg. The work felt important. She also had personal reasons to return to the West Coast. Her sister, Michelle, lived in San Francisco, and Sandberg had many friends in Los Angeles. She began to date one of those friends, Dave Goldberg, a music tech start-up founder in 2002. They were engaged six months later. After he sold his company, Launch, to Yahoo in June 2003, Goldberg stayed in LA at first, but joined Sandberg in the Bay Area a year later. "I lost the coin flip as to where we were going to live," he said. In 2004, they were married in a desert wedding in Carefree, Arizona.

Sandberg thrived at Google and was credited with growing the search engine's nascent advertising business into a $16.6 billion enterprise. She was quoted in the *New York Times*, *The New Yorker*, and *Newsweek*, and spoke at conferences like the *Wall Street Journal*'s AllThingsD with Kara Swisher and Walt Mossberg. Start-up founders got headlines, but those in the know recognized that it was people like Sandberg who worked behind the scenes to turn early-stage start-ups into Fortune 100 corporations.

She met Zuckerberg at a Christmas party thrown by former Yahoo executive Dan Rosensweig in December 2007 at his home in Woodside. Zuckerberg was avoiding small talk, but he was interested in talking to Sandberg. As guests milled around them socializing, the two discussed business. He described his goal of turning every person in the country with an internet connection into a Facebook user. It might have sounded like a fantasy to others, but

Sandberg was intrigued and threw out ideas about what it would take to build a business to keep up with that kind of growth. "It was actually smart. It was substantive," Zuckerberg later recalled.

That they would cross paths was inevitable, given their many connections. Rosensweig was one of several friends Zuckerberg and Sandberg had in common. Goldberg had worked with Rosensweig at Yahoo; Rosensweig had shaken hands with Zuckerberg over the failed Yahoo deal. Roger McNamee was an early investor in Facebook, and his partner at his investment firm, Elevation Partners, was Marc Bodnick, who was married to Sandberg's sister, Michelle.

Zuckerberg and Sandberg ended up talking for more than an hour at that party, standing in the same spot near the entrance. She was intrigued by the guy who had turned down Yahoo's $1 billion buyout offer and by the nerve it took to walk away in the face of employee protests in support of the deal. And she found his ambition of exponentially increasing Facebook's fifty million users irresistible. As she would later tell Dan Rose, a former vice president at Facebook, she felt that she was "put on this planet to scale organizations." Facebook's earliest business executives, including Sean Parker, didn't know how to run or grow a business at global scale.

After the Christmas party, Zuckerberg and Sandberg met at her favorite neighborhood restaurant, Flea Street, and over several long dinners at her home in Atherton. As their discussions grew more serious, they hid their meetings to avoid speculation about her departure from Google. They couldn't meet at Zuckerberg's apartment, a studio in Palo Alto with only a futon mattress on the floor, a small table, and two chairs. (In 2007, Google's founders, Brin and Page, and Chief Executive Eric Schmidt discussed a partnership with Zuckerberg in the tiny apartment. Zuckerberg and Page sat at the table, Brin sat on the futon, and Schmidt sat on the floor.)

In their conversations, Zuckerberg laid out his vision for what he called "the social web," how Facebook was an entirely new communications technology, one where news entertainment would be generated for free by its users. Sandberg walked Zuckerberg through how she had helped scale Google's ad business, turning search queries into data that gave advertisers rich insights about users, contributing to the company's spectacular cash flow. She explained how Facebook would increase its number of employees, budgeting more money for capital expenses like data centers, and grow revenues at a manageable pace to keep up with the explosion of new users.

Their conversations ran so long that Sandberg, who had a newborn and a toddler, had to kick Zuckerberg out so she could get some sleep.

In some ways, they were opposites. Sandberg was a master manager and delegator. At Google, her calendar was scheduled to the minute. Meetings rarely ran long and typically culminated in action items. At thirty-eight, she was fifteen years older than Zuckerberg, was in bed by 9:30 p.m. and up every morning by 6 for a hard cardio workout. Zuckerberg was still dating his Harvard girlfriend, Priscilla Chan, who had just graduated and was working at a private school in San Jose, thirty minutes away from Palo Alto. But he was focused mostly on his work. He was a night crawler, coding way past midnight and up in time to straggle into the office late in the morning. Dan Rose recalled being pulled into meetings at 11 p.m., the middle of Zuckerberg's workday. Sandberg was hyperorganized and took voluminous notes in a spiral-bound scheduler, the kind she had used since college. He carried around his laptop and would show up late to meetings, or not at all if he was in a conversation or a coding groove he found more interesting.

Zuckerberg recognized that Sandberg excelled at, even enjoyed, all the parts of running a company that he found unfulfilling. She had high-level contacts at the biggest advertising agencies and with the top executives of Fortune 500 companies. And she would bring to Facebook an asset that her new boss knew he would need: experience in Washington, DC.

Zuckerberg wasn't interested in politics and didn't keep up with the news. The year before, while shadowing Donald Graham for a few days at the *Post*, a reporter handed him a book on politics that he had written. Zuckerberg said to Graham, "I'm never going to have time to read this."

"I teased him because there were very few things where you'll find unanimity about, and one of those things is that reading books is a good way to learn. There is no dissent on that point," Graham said. "Mark eventually came to agree with me on that, and like everything he did, he picked it up very quickly and became a tremendous reader."

Yet, in the lead-up to his talks with Sandberg, Zuckerberg had had a brush with controversy that stoked concerns about potential regulations. Government officials were beginning to question if free platforms like Facebook were harming users with the data they collected. In December 2007, the Federal Trade Commission issued self-regulatory principles for behavioral advertising to protect data privacy. Zuckerberg needed help navigating Washington. "Mark understood that some of the biggest challenges Facebook was going to face in the future were going to revolve around issues of privacy and regulatory concerns," Rose said. "[Sandberg] obviously had deep experience there, and this was very important to Mark."

To Sandberg, the move to Facebook, a company led by an awkward twenty-three-year-old college dropout, wasn't as counterintuitive as it might have appeared. She was a vice president at Google, but she had hit a ceiling: there were several vice presidents at her level, and they were all competing for promotions, and Eric Schmidt wasn't looking for a number two. Men who weren't performing as well as she were getting recognized and receiving higher titles, former Google colleagues maintained. "Despite leading a bigger, more profitable, faster-growing business than the men who were her peers, she was not given the title president, but they were," recalled Kim Scott, a leader in the ad sales division. "It was bullshit."

Sandberg was looking for a new endeavor. She'd been courted for multiple jobs, including a senior executive role at the *Washington Post*. Don Graham had struck up a relationship with her when she was working for Summers, and in January 2001 he had tried to recruit her during a luncheon with his mother, Katharine, but she chose to join Google instead. Six years later, he tried to recruit her again. He was nervous about the decline in print news publishing and offered her a role as a senior executive and the chance to turn an old media brand into an online powerhouse. Again, she declined.

In spite of the fact that he had been rejected by both Zuckerberg and Sandberg, Don Graham remained close with them, and they continued to seek his counsel. When each asked him for his opinion about the other, he encouraged them to join forces. Toward the end of January 2008, less than a month after they first met, Zuckerberg accompanied Sandberg and other executives on Google's corporate jet to the World Economic Forum in Davos, and over several days in the Swiss Alps, they continued to discuss a vision for Facebook. On March 4, Facebook sent out a press release naming Sandberg as chief operating officer of the company.

The headline the *Wall Street Journal* gave its coverage of the hire, "Facebook CEO Seeks Help as Site Grows Up," spelled out Sandberg's challenge. She was in charge of operations, growing revenues, and global expansion. She was also in charge of sales, business development, public policy, and communications.

"Mr. Zuckerberg and Ms. Sandberg will face mounting pressure to find a better business model," the article read. "Facebook's Web traffic continues to rise. But industry watchers are now questioning whether that growth will ever translate into Google-size revenue."

At the time of Sandberg's appointment, the company's four hundred employees occupied several offices around University Avenue, an upscale thoroughfare of boutiques and restaurants in Palo Alto that was within walking distance of the juniper- and palm-lined campus of Stanford University. Apple CEO Steve Jobs lived nearby and was sometimes seen walking in the neighborhood in his trademark black turtleneck and jeans.

The office culture hadn't changed much since the company's earliest days. Half-consumed bottles of Gatorade and Snapple populated workstations that held stacks of coding books. Employees rode scooters and RipStiks down the halls. Achievements were invariably celebrated with booze drunk from Red SOLO cups.

Plenty of other start-ups offered more perks. But these employees had chosen Facebook in large part because of Zuckerberg himself. He was one of them, a "product guy" who knew how to make things. It was an intensely competitive environment. (On a whiteboard at the Emerson Street office, an engineer had jokingly kept a tally of the number of times he needed to go to the bathroom but had resisted, a barometer of his focus and commitment.) Engineers placed bets on how late into the night they could code complex projects.

One engineer who helped set the tone was Andrew Bosworth, a Harvard grad whom Zuckerberg hired in 2006 to help lead engineering and who was one of the creators of the News Feed. Bosworth, known to all as Boz, was an intimidating presence, fully bald and with the stature of a linebacker. He was a computer science TA at Harvard when he met Zuckerberg in an Artificial Intelligence course; the two kept in touch after Zuckerberg didn't return for his junior year. Boz was a jokester and full of energy, with no filter on his thoughts. He drove the engineers hard and embraced a tough-guy image. He had a joke catchphrase, "I'll punch you in the face," which colleagues grew accustomed to hearing. It was emblazoned on a T-shirt he would wear every so often. Some employees complained, and at the end of 2007, in a winding Facebook post, he announced that he was retiring it: "I enjoyed my moment in the sun as a nominal part of the budding Facebook culture," he wrote. "Unfortunately, it has come to my attention that as a loud, 6'3", 250lb, often bearded beast of man I am not well served by an intimidating reputation."

It was a difficult place for Facebook's few female employees. Katherine Losse, employee number fifty-one and eventually Zuckerberg's speechwriter, would later recall the demeaning comments casually made about women around the office. One day, speaking to a female colleague, a male coworker said, "I want to put my teeth in your ass." At a company meeting in which the incident was raised, Zuckerberg attempted to dismiss the comment, asking, "What does that even mean?" Later, when Losse followed up with Zuckerberg, she was struck by his callowness. "He listened to me, which I appreciated, but understanding the crux of the matter; that is, that women by virtue of our low rank and small numbers were already in a vulnerable situation in the office, did not seem to register."

On Sandberg's first day, Zuckerberg called an all-employee meeting and introduced the COO as the person who was going to help him "scale" the company. Sandberg gave a short speech explaining her role.

"Who's Boz?" she then asked the crowd. She heard he had written a post expressing concern that the company was growing too quickly. He was afraid Facebook would lose its hungry hacker culture.

Clearly surprised, Boz sheepishly raised his hand. She thanked him. His post had helped her decide to join the company, she said. Her responsibility was to grow the company while retaining what made it special. "We're going to have one thousand people someday, and we're going to have ten thousand people someday, and then forty thousand people someday," she said. "And we're going to get better, not worse. That's why I'm here. To make us bigger and better, not worse."

It was a good pep talk. The mostly male room of engineers warmed up. She was coming in to steer the business safely, not change its direction. It had been traveling fast, and she was there to make sure it became successful and not just another tech start-up with fifteen minutes of hype.

Zuckerberg, for his part, wasn't as eloquent. He mentioned to staff that Sandberg had "good skin," and said they should have "a crush" on her, according to Losse.

Sandberg was like an alien to the young, upstart staff: she'd been a North Miami Beach High School student with big hair, big shoulder pads, and creamy blue eyeshadow when Zuckerberg, and most of his employees, were still in diapers. And on the surface at least, it was a strange landing for her. For much of her adult life, she had advocated for gender equity. Her college thesis examined how women in abusive domestic relationships were more likely to suffer financially. As a young associate at McKinsey, she stood up to a client who harassed her about dating his son. And at Google, she started Women@Google, a speakers' series that brought Jane Fonda, Gloria Steinem, and other prominent women leaders to the company's Mountain View campus.

Soon after she joined, Sandberg remarked that Facebook felt like the early days of Google. She didn't comment on the male-driven culture, but employees like Naomi Gleit, one of the first women hired at the company, said she was relieved to see another woman in charge. Sandberg had gone to nearly every desk to introduce herself in the early days and made a particular point of spending time with female employees.

But not every female employee found herself in Sandberg's favor. "It was very clear, starting in the second month or so, whom she was most interested in helping. They were the ones that looked like her and came to work with their gym bags or yoga bags and blownout hair," said one early employee. "They were not scruffy engineers who were in the trenches daily on male-dominated teams."

A month into her new job, Sandberg gathered half a dozen members of the ad team for a working dinner in one of the Palo Alto office conference rooms. In attendance were Kendall and Rose, who was managing a partnership Facebook held with Microsoft for display ads. Chamath Palihapitiya, the head of growth; Kang-Xing Jin, a software engineer whom Zuckerberg had met at Harvard; and Matt Cohler, the head of product management, rounded out the group.

Zuckerberg had gone on a rare extended vacation. The site had become global and had just launched in Spanish, with 2.8 million users in Latin America and Spain, but he himself had limited life experience and exposure outside his bubbles of Dobbs Ferry, Exeter, Harvard, and Silicon Valley. Steve Jobs, who had become a mentor or sorts, taking Zuckerberg on a few long walks in the hills behind Stanford University, encouraged him to see the world. Zuckerberg set off on a one-month trip, with stops in Germany, Turkey, Japan, and India, where he spent time at an ashram that Jobs had suggested he visit.

The trip was perfectly timed, affording Sandberg the opportunity to plot out the reinvention of the revenue-generating part of the company. "What business are we in?" she asked the leaders while scribbling furiously on a whiteboard. A subscription business or an advertising business? Did they want to make money by selling data through payments or through commerce?

There wasn't much deliberation. Of course Facebook would be free, the executives said, so their best path to make money was through ads.

Sandberg nodded. To do this well, she continued, they would have to pinpoint what Facebook really had to offer. The difference between Google and Facebook, she continued, was the kind of data the two companies collected. Employing a well-known metaphor used by MBAs and Madison Avenue types, she described reaching customers through advertising as a funnel. Google, she explained, was at the bottom of the funnel, the narrow neck at which consumers are closest to pulling out their credit cards. Google used data from search queries to push over the finish line those users already in the mood to shop. Google didn't really care who the individual was. It cared only about the key words entered into its search bar so it could churn out ads targeted for the search results.

The meeting was an exercise in the elementary aspects of advertising, but she wanted to start it with them all on the same page.

Just as the internet was opening up entire new worlds of information and connection for its users, it was also opening up a whole new world of promise for advertisers, she continued. The question was, how best to capture this audience? The visual real estate was entirely different from print, and so was viewers' attention span.

In 1994, an engineer at Netscape had created the "cookie," a string of code dropped by the Web browser to track users as they surfed sites across the internet, data that was then sold to advertisers. The cookie opened up the possibility of deeper tracking called "behavioral advertising" and put a new emphasis on the importance of data collecting. But for more than a decade, internet advertising was dominated by static, uninventive postage stamp–size ads, banner ads, and obnoxious pop-ups. Advertisers didn't see a whole lot of return despite the ability to track users.

Many of the people in the conference room that day had been working on Facebook's early ad targeting, which had started soon after the platform got off the ground. But their actual ads hadn't developed beyond banner ads and the occasional sponsored link; internet users were ignoring them, studies showed. Much of the ad business was essentially outsourced to Microsoft, which in 2006 struck up a deal with Facebook to sell and place banner ads on the social network.

Sandberg had been in charge of Google's in-house ad auction known as "AdWords" and of another program, called "AdSense," which was used by publishers to run text and graphic ads on sites across the internet. The biggest innovation was Google's use of search data to spit out ads directly related to a user's query. Type in "Hawaii flights cheap," and an auction immediately ran behind the scene for vendors to bid to place their ads for Honolulu budget hotels, swimwear, and surfing lessons within search results. It was the purest and most powerful way of directly pitching products to someone already in the mood for shopping.

AdWords and AdSense had turned Google into an advertising behemoth. And Sandberg was expected to apply her expertise to jump-start Facebook's ad business, which was lagging in proportion to its skyrocketing user growth. The executives gathered with Sandberg that day were there to diagnose what wasn't working and to home in on Facebook's unique proposition to advertisers.

Over the course of a few more meetings, the group determined that Facebook would focus on the biggest and most traditional form of advertising: brand awareness. Facebook, Sandberg pointed out, had a distinct advantage over television, newspapers, and websites like Yahoo and MSN, which had dominated brand advertising in recent years. The social network had much more than just detailed profile information on users; the company also had information about everyone's *activity*. Facebook's great strength was the unparalleled engagement of its users. Facebook knew its users, and it had all their data in one place, so that advertisers didn't have to rely on cookies.

A forty-year-old mother living in a wealthy suburb who had visited a Facebook page of a ski resort in Park City, Utah, for instance, and who had other wealthy suburban friends who shared photos of their ski vacations, was likely the perfect audience for UGG's $150 shearling-lined boots. And rather than sell ugly banner ads, Facebook could invite brands to create, and capitalize on, user activity by designing campaigns on the site that drew users to interact through polls, quizzes, and brand pages. Users themselves could comment and share these with friends, generating even more data for both Facebook and its clients.

The way Sandberg saw it, if Google filled demand, then Facebook would create it. Facebook users weren't there to shop, but advertisers could use the company's understanding of its users to convert them into shoppers. What she was proposing was essentially a whole new level of tracking. The plan to execute her vision was similarly bold, involving hiring more staff to work under her as well as engineers to develop new ad products. It required Facebook to participate in ad shows and to court the biggest brands.

When Zuckerberg returned from his vacation, he signed off on the new business model. Facebook would run its own ad business and, eventually, rely less on Microsoft as it focused on targeted advertising and new ad tools like games and promotions to get brands directly in front of users. But he wasn't all that interested in the details—an attitude that Sandberg found baffling. She had thought, from his interest during their courtship dinners, that Zuckerberg was committed to strengthening the business and that he would be actively supportive of her role. As the weeks went by, though, it became apparent that she would have to fight to capture his attention. When she suggested carving out time to talk about the ad business, or asked for more staff or a bigger budget, he brushed off her requests. It was clear that profits ranked far below technology. This was something all Zuckerberg's early employees knew. "His first priority will always be to build a product that people like, that they feel good about, and that fills a particular need," explained Mike Hoefflinger, who was the head of business marketing in 2008.

When Sandberg was hired, she had insisted on twice-weekly meetings with Zuckerberg to talk about their agendas and the biggest problems facing the company. She scheduled the meetings for Monday and Friday, so they could start and end the week together. But she didn't naturally command his attention inside and outside the office in the way Boz and Cox, his friends and peers, did. She struggled to get the resources and the attention she wanted to grow the ad business more quickly. A bifurcation, a wedge between the teams of tech and non-tech employees, soon developed. The people she poached from Google and hired from her network of Treasury alumni became known as "Friends of Sheryl Sandberg," or FOSS. Most senior employees were unofficially divided into two camps, "Sheryl people" or "Mark people."

It could get territorial. Sandberg gave unsolicited opinions on products and engineering, some employees said. Ad tools required developers to build them, but Zuckerberg was reluctant to take engineers off the development of new features for the site to help build Sandberg's side of the ad business. There was mutual distrust and annoyance. Engineers looked down on the "marketing guys," and the business managers struggled to get into meetings with Zuckerberg's teams. "I've been consistently puzzled about the public portrayal of them as amazing partners," one business-side employee observed. "Meetings I've had with Mark and Sheryl, he was explicitly dismissive of what she said."

Sandberg had begun at Google in the same place: with few resources and little attention from founders who held their nose at making money. Undeterred, she began wooing the biggest brands, including Ford, Coke, and Starbucks. In her pitches, she emphasized how, unlike users of other platforms, including the rapidly growing Twitter, users on Facebook were required to give their real identities. Sandberg asked: Wouldn't you want to be part of those conversations? Facebook was the biggest word-of-mouth advertising platform in the world, she said. With the right campaigns, users themselves would persuade their friends to buy products. Meanwhile, the brands could keep track of users who made comments about them and joined their brand pages.

Even with Sandberg's reputation, the major brands were hard to convince. The heads of the top ad agencies doubted that people coming to the platform to comment on vacation photos would willingly spread the word about Verizon wireless service. And while her message about creating demand resonated with advertisers, they weren't convinced that Facebook could influence consumer hearts and minds better than a well-placed primetime television spot. Its targeting abilities may have been enticing, but the social network was still using display and banner ads that didn't seem to offer anything new.

Sandberg was responsive to the feedback from Facebook's advertising skeptics. When Sony's chief executive, Michael Lynton, expressed his doubt about the payoff of advertising on a platform that wasn't even measured by the ad industry, Sandberg responded by negotiating a partnership with Nielsen, the media audience measurement agency, to measure attention for ads on the social network. "Unlike any other executive, instead of just nodding her head, she and Facebook acted incredibly quickly," Lynton said.

She was making headway, but was thwarted by Zuckerberg. She

needed his participation in these sales pitches. She needed more advertising sales staff, and engineers to design new ad features. Instead, he was hoarding the engineering talent to work on user features. She finally took her complaints to Don Graham, who in late 2008 was named to Facebook's board of directors, which had expanded to include Marc Andreessen. She asked Graham to intervene and lobby Zuckerberg for more resources on her behalf.

As the only board member who worked in a business that also relied on advertising, Graham understood Sandberg's frustration, but it was hard to convince Zuckerberg to take her concerns seriously. Graham was also mindful of the limits of his role; the company was ultimately Zuckerberg's. "His priority of growth was miles above advertising, revenue, and profits," Graham recalled. "It became my job to call him after every time Sheryl called and try to convince him about the importance of advertising, and not to change his priorities, but to help move him a little."

Well before Sandberg arrived at Facebook with a vision for ramping up its data mining efforts, Jeff Chester had been following Facebook's growth warily. In November 2007, the privacy rights advocate had been sitting in his tiny office space in Washington, DC, following a live blog of Zuckerberg on his laptop. The Facebook founder, addressing ad executives in New York City, was announcing the company's latest innovation: a revolutionary program called "Beacon."

Chester stiffened. Beacon, Zuckerberg continued, was an ad initiative that took information about a Facebook user's purchases from other sites—movie tickets on Fandango, a hotel booking on Tripadvisor, a sofa on Overstock—and published it on the News Feeds of his or her friends. The companies were partners in Beacon, eager to be part of a program that gave them data on Facebook users and also promoted their goods and services through recommendations—even if they were involuntary. A purchase made, story read, or recipe reviewed would automatically show up in user feeds. The ideal in advertising was word-of-mouth endorsements, and Facebook was offering a way to provide them on a massive scale: their users would effectively serve as brand evangelists. Beacon erased the line between advertising and "organic" user comments. "Nothing influences a person more than a recommendation from a trusted friend," Zuckerberg said. Already, Facebook had signed more than forty partners—including CBS, the *New York Times*, and TheKnot—that were paying Facebook for the privilege of getting their brands in front of the social network's users.

Chester jumped out of his chair and called his wife, Kathryn Montgomery, a professor of media studies at American University. "This is unbelievable!" he shouted. "You've got to hear what this guy from Facebook is saying!" From Chester's perspective, Facebook's plan was the logical extension of what advertising had always done: hijack the minds of consumers to persuade them at the checkout counter. Chester, for his part, had been fighting what he saw as manipulation since the 1990s, when he railed against television broadcasters and advertisers for product placement on shows and the promotion of junk food on children's programs. With the advent of the internet, he had trained his sights on the unregulated world of online advertising. He'd founded a privacy rights group called "the Center for Digital Democracy," and in 1998, with the help of Montgomery, he'd successfully pushed for a law protecting kids on the internet, known as the federal Children's Online Privacy Protection Act.

That the Center for Digital Democracy consisted of just one full-time employee, Chester, and was located in Dupont Circle, far from the center of political activity near Capitol Hill, did not discourage him. The native Brooklynite relished the role of outsider, and with his rumpled slacks, rimless glasses, and disheveled hair, he looked the part, too, a standout in the slick suit-and-tie universe of Washington lobbyists. He scoured about ten technology and advertising industry trade publications every day, looking to uncover unsavory business practices, which he then compiled into individual emails to journalists. His one-liners often proved irresistible. Just a few days before Zuckerberg's announcement, Chester had been quoted in the *New York Times* describing behavioral advertising as a "digital data vacuum cleaner on steroids."

In his mind, the Beacon announcement marked a dangerous new low. Zuckerberg had not asked permission from Facebook account holders to use them as sales agents; Beacon enrolled them automatically. Facebook was widening its data net, exploiting insights about its users in ways that crossed ethical lines. He called privacy groups, fired off a statement to the press, and reached out to his media contacts.

"It was a wakeup call. Beacon had the seeds of all the problems that would come later," Chester recalled. "Regardless whether the user wanted to or not, Facebook was going to monetize everyone who was using the social network and turn individuals into advertisers."

The next day, Facebook users were shocked to see their private lives on display. Suddenly, that Travelocity flight reservation to Florida, last night's guilty pleasure movie rental, or eBay bid was popping up in the News Feeds of friends, relatives, and coworkers. One woman discovered that her boyfriend had bought her a diamond ring.

The outcry was immediate. On November 20, the public advocacy organization MoveOn.org circulated a petition asking Facebook to shut down the service; within days, it had amassed fifty thousand signatures. As the controversy gathered momentum, Coca-Cola and Overstock dropped out of the program, telling reporters they had been misled into believing that Beacon would require user activation. When Facebook responded that the feature could be turned off, users produced contrary evidence. A security researcher at Computer Associates said that after he had opted out of Beacon, he noticed network traffic patterns revealing that recipes he had saved on Epicurious were tracked by Facebook.

The public fury over Beacon was missing the point, Chester thought. Sure, it was embarrassing and invasive to have your shopping activity exposed to acquaintances and family members. But that was a distraction from the real threat: the fact that they were being tracked and monitored by an entirely new entity. Everyone knew to be wary of the government's reach, but in Chester's estimation the danger wasn't what the public or law enforcement knew about you. It was what commercial enterprises and advertisers did. It was what Facebook knew.

Mere weeks after his appearance in New York, Zuckerberg apologized for springing the new ad tool on users and announced that he would change the setting to make it an opt-in instead of a default program requiring users to opt out. In a blog post, he assured users that Facebook would stop sharing their shopping activity without permission. He acknowledged that the rollout had been botched. "We missed," he said, "the right balance."

The controversy, and Zuckerberg's weak apology, completely ignored the real privacy abuses. As Chester and Montgomery saw it, personal data should be kept out of view from advertisers—and Facebook hadn't done anything to change that. "Facebook said you can choose whom you share information with on the platform," explained Montgomery, "but behind the platform, where you don't see what is really happening and how they are making money, they don't give you a choice about what the platform shares about you with advertisers."

With Sandberg's hiring, the company entered a new phase of advertising. She lured big brands like Adidas and Papa John's Pizza to create quizzes and fan pages to get users to engage directly with advertisers and was overseeing the development of ad targeting based on geography and languages.

She also swiftly emerged as the most effective spokesperson for the company, fulfilling one of the key roles the CEO had in mind for his number two. She crafted an entirely new spin on Facebook's handling of data privacy and repositioned Facebook as a leader on the issue, pointing to how it offered users granular controls over who (the public, friends, selected individuals) could see particular content. Facebook didn't "share" data with advertisers, she asserted, a talking point the company would repeatedly fall back on, even though some critics argued it was a distinction without a difference. True, Facebook didn't physically hand over or directly sell data to advertisers. But advertisers were targeting users by age, income, employment, education, and other demographics. As Siva Vaidhyanathan, a professor of media studies at the University of Virginia, pointed out, the evidence was clear that the company's profits came from data: "Facebook has been duplicitous and irresponsible with our data for years."

Sandberg insisted that Facebook was putting power in the hands of its users to express themselves without fear. "What we believe we've done is we've taken the power of real trust, real user privacy controls, and made it possible for people to be their authentic selves online," she said at a tech conference in November 2008. "And that is what we think explains our growth."

The statements were in direct contradiction to what privacy advocates were seeing. The company's profits, after all, were contingent on the public's cluelessness. As Harvard Business School professor Shoshana Zuboff put it, Facebook's success "depends upon one-way-mirror operations engineered for our ignorance and wrapped in a fog of misdirection, euphemism and mendacity."

Silicon Valley's other tech executives seemed only too happy to perpetuate this ignorance. ("If you have something that you don't want anyone to know about, maybe you shouldn't be doing it in the first place," Sandberg's former boss Eric Schmidt would famously quip in a 2009 interview on CNBC, echoing the law enforcement refrain to emphasize user responsibility.) And in fact, Facebook was about to turbo-charge its data collection operation. In February 2009, it unveiled the ultimate vehicle for quick and free expression: the Like button. The feature, which would come to define not just Facebook but nearly every exchange on the internet, had been designed by one of the company's product managers, Leah Pearlman. She began working with Boz and other senior Facebook executives on her idea in 2007, but the project stalled because of staff disagreements: over what the button should be called, whether there should also be a Dislike button, and whether any such quick-response button would diminish engagement minutes.

Facebook had created a culture where engineers were encouraged to ship, or deliver, products as quickly as possible. "Fuck it, ship it," was a popular expression used across the company. But in this case, the project inched forward at a snail's pace. In late 2008, Zuckerberg finally gave his blessing. Internal data had convinced him of the feature's worth: in small tests, people used Facebook more when the button was available. He ruled unilaterally on the official name: "the Like button."

It was an immediate hit. As users scrolled through their Facebook News Feed, the button gave them a way to send quick, positive affirmation to their friends. If you liked an item on News Feed, Facebook showed you other, similar content, which meant that suddenly you could scroll through an endless, and endlessly hilarious, stream of cat videos and funny memes. Meanwhile, in their personal posts, users began competing for likes, sharing more and more of themselves to accumulate digital thumbs-ups. Unlike

Beacon, which Facebook discontinued that same year, the Like button met with little resistance. Pearlman had unknowingly designed a new currency for the internet, one through which politicians, brands, and friends competed for validation.

The button was so successful that, nearly a year after its launch, Facebook decided to make the feature available outside its site. Any website could now add the Facebook Like button simply by inserting a tiny line of code. It was a win-win situation: companies got information about which Facebook users were visiting their sites, and Facebook received information about what its users did once they left its site.

To Facebook users, the Like button was useful. Because of it, Facebook could show them the pages, groups, and interest pages they were more likely to join on Facebook. Facebook could also see what their friends were liking outside Facebook and suggest the same thing to users it believed were like-minded and shared interests.

But within Facebook itself, the Like button was more than useful. The feature represented an entirely new capability and scale of collecting insights into users' preferences.

By the end of 2009, Facebook had 350 million users. The company was facing competition, however: its rival Twitter was gaining both popularity and buzz. With 58 million users, Twitter was much smaller than Facebook, but Zuckerberg worried about a key advantage the site had: Twitter was public.

This meant a Twitter user could look up any other user and follow their account. Status updates were available to anyone on the site. Facebook, however, was organized around closed networks. Users decided if they wanted to accept friend requests, and they could hide most of their profile data to evade a simple Web search. Facebook was a gathering place for closed-door discussions among friends. Twitter was a loud, noisy, and increasingly crowded town square. Katy Perry and the Dalai Lama had joined the platform, instantly acquiring millions of followers.

Facebook needed to capture some of the same excitement. Zuckerberg was fixated on the new rival. He ordered staff to come up with a response. In December 2009, he announced a gutsy move: certain user details previously set to "private" were being switched to "public." That same day, when users logged on to Facebook, they were greeted with a pop-up box asking them to retain the "everyone" setting, the most public option. But the request was confusing, and many users clicked to consent without understanding the implications. Previously hidden personal information (photos, email addresses, and more) was now searchable on the Web. Users started receiving friend requests from strangers. The new settings were cumbersome and opaque; switching back to more private modes seemed especially tricky. The change was also controversial within the company. Before the launch, one employee on the policy team met with Zuckerberg and told him that the feature would become a privacy disaster. Consumers don't like to be caught off guard, and if they feel hoodwinked by the change, Facebook will attract unwanted attention in Washington, the employee warned. But Zuckerberg's mind was made up.

As expected, the change in privacy settings set off a storm of anger. Facebook claimed the changes were intended to streamline its confusing system of privacy settings. But really, it was opening up the network in an effort to become a center of activity on the internet. As *TechCrunch* wrote, "In short, this is Facebook's answer to Twitter. . . . That means Facebook can leverage [public profiles] for real-time search, and can also syndicate it to other places, like Google and Bing."

That same week in Washington, regulators attending the annual meeting for the International Association of Privacy Professionals pressed Facebook's lobbyist Tim Sparapani about the changes. Sparapani defended the action as better for privacy, touting the ability of users to pick through a detailed menu of controls. But government officials were reading the news coverage. Some stories quoted Jeff Chester, who characterized the public exposure of user profiles as illegal because it seemed to violate consumer fraud and deception laws. Facebook's definition of privacy is "self-serving and narrow," Chester asserted. "They don't disclose that consumer data is being used for very sophisticated marketing and targeting."

Zuckerberg didn't seem to understand why users were so outraged. Unlike tech CEOs from earlier generations, such as Bill Gates, Larry Page, and Sergey Brin, who fiercely guarded their privacy, he projected a carefree attitude about sharing information online. On the public settings for his own Facebook page, he posted photos of himself with Priscilla and happy hours with friends and colleagues. The implicit message seemed to be that he had nothing to hide. He didn't see the problem with collecting more data and exposing user profile information; it felt like a small price to pay for free connection to anyone in the world. (In January 2010, he went so far as to declare in a speech at *TechCrunch*'s Crunchies awards that sharing online was becoming a "social norm.")

He couldn't recognize, perhaps, that his life experience—his safe and stable upbringing, his Ivy League pedigree, his ability to attract investors—was not the same as everyone else's. As with his lack of empathy for his female employees, he couldn't identify the systemic biases of the world: how, if you were Black, you might attract ads for predatory loans; or if you were lower-income, you might attract ads for junk food and soda.

But it was clear he understood the personal risks of sharing too much. While he preached openness to the public, he was protective of his Facebook account and curated which friends were granted full access to his photos and posts. His life off-line was even more guarded. (In 2013, he bought the houses around his home in Palo Alto and tore them down to expand his lot.)

Earlier that year, when journalist Jose Antonio Vargas had interviewed Zuckerberg for a story in *The New Yorker*, he'd challenged the tech wizard on his notions of privacy. Sitting on a bench outside the small rental home Zuckerberg shared with Priscilla, Vargas shared that he was gay but hadn't come out to his relatives in the Philippines. The secret could pose great risks to him if exposed. But Zuckerberg didn't seem to grasp the issue and was at a loss for words, staring at Vargas with a blank expression, a pause "so pregnant," Vargas recalled, that "it gave birth." In Zuckerberg's worldview, the authenticity of individuals was the most valuable proposition—to users and advertisers alike. His reaction also spoke to his limited, sheltered life experience. "If your life is shaped by Westchester, Exeter, and Harvard, and then you move to Silicon Valley, if that's your understanding of the world, then that's a very specific world," Vargas added.

Facebook's cavalier stance toward user data was also rankling government officials in Washington. In December 2009, aides to Sen. Chuck Schumer of New York complained to Facebook's lobbyist Sparapani about the change to privacy settings, which they were struggling to reset to "private" on their own accounts. The call from Schumer's office spooked the company and prompted a visit by Sparapani and Elliot Schrage from Palo Alto. Schrage was a Sandberg hire, a confidant from Google who served, officially, as Facebook's head of policy and communications and, informally, as Sandberg's consigliere. Sparapani, the company's first registered lobbyist, had been poached from the ACLU in anticipation of the coming storm over personal data.

In Schumer's office, the pugnacious Schrage did much of the talking. He was impatient with questions from Schumer aides, who complained that the company seemed to have made it harder to make content private. Schrage said the new settings weren't problematic, repeating Sandberg's message that Facebook had the strongest privacy policies of any internet company. His dismissive posture irritated Schumer's aides. The confrontational style and defensiveness of the visitors from the tech company frustrated the congressional staff in the room. Schrage seemed to be setting the tone for the platform's approach to government: Facebook didn't seem to be listening to their concerns; nor did it intend to, staffers recalled.

In the wake of the meeting with Schumer and the consumer uproar, Facebook dialed back some of the privacy options for users so that most data, except for names and profile pictures, was no longer made public without permission. But the damage had been done. Regulators started to take notice of privacy issues. Jonathan Leibowitz, appointed by President Obama as head of the FTC earlier in the year, proved a prescient choice. Leibowitz, a veteran congressional aide and former lobbyist for the Motion Picture Association of America, had famously given a harsh review of the internet's nascent behavioral advertising industry, saying he was "troubled" by some companies' "unfettered collection and use of consumers' sensitive data," especially information about children and adolescents. "The possibility that companies could be selling personal identifiable behavior data," he noted, was "not only unanticipated by consumers, but also potentially illegal."

On December 17, 2009, the Center for Digital Democracy and nine other privacy groups filed a complaint at the Federal Trade Commission charging that Facebook's change to its privacy settings was illegally deceptive. The complaint was aimed at the agency's obligation to protect consumers from deceptive and unfair business practices.

In January 2010, the FTC responded to the CDD's complaint, noting that it was of "particular interest" and "raises a number of concerns about Facebook's information sharing practices." The letter was highly unusual; the commission rarely signaled its interest in a case. Facebook was, for the first time, the subject of a major federal investigation.

That investigation would result in a settlement that subjected the company to regular privacy audits for two decades. And yet, over the next few years, the government would largely leave Facebook to grow, opting not to stand in the way of the Instagram and WhatsApp mergers of 2012 and 2014 and offering little oversight. It wasn't until Facebook entered into its period of crisis in 2016 that the ramifications of the FTC's actions would once again come into play.

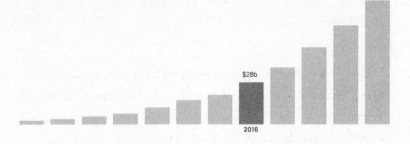

$28b

2016

Chapter 4

The Rat Catcher

They called her "the rat catcher."

In the winter of 2016, employee morale at Facebook was flagging. Websites posting false information and partisan conspiracy theories were regularly among the top ten most-viewed sites on the platform. Employees complained to their managers that the company was spreading harmful information. In their various Tribe boards, engineers asked whether their algorithms favoring sensational material made the network appear to be a tool of Donald J. Trump's presidential campaign.

And for nearly six months, some employees shared their grievances with reporters, passing on internal memos, speeches, and emails. Executives were furious; Boz in particular ranted about how Facebook was a trusted family and that a small group of outsiders was corrupting and corroding its values. After a story headlined "Mark Zuckerberg Asks Racist Facebook Employees to Stop Crossing Out Black Lives Matter Slogans" appeared on the tech site Gizmodo that February, the senior leaders turned to Sonya Ahuja, a former engineer whose internal investigations unit hunted down everything from harassment and discrimination complaints to leakers and whistleblowers.

Ahuja's department had an eagle's-eye view into the daily work-ings of Facebook employees, with access to all communications. Ev-ery move of a mouse and every tap of a key was recorded and could be searched by Ahuja and her team. This kind of surveillance was built into Facebook's systems: Zuckerberg was highly controlling of Facebook's internal conversations, projects, and organizational changes. When a product was announced before he intended or when internal conversations were reported in the press, the security team was deployed to root out the traitor within. Facebook's secu-rity team also tested employees by leaving "mousetraps" around the company, with what appeared to be secret information, to see if employees would turn them in. While it had become common-place for Silicon Valley companies to demand nondisclosure agree-ments from employees, which barred them from discussing work with anyone outside the company, Facebook demanded that many employees sign an additional nondisclosure agreement. Facebook employees could not even talk about the things they were not al-lowed to talk about. Even by the standards set by Google, Twitter, and other Silicon Valley social media companies, Facebook was notorious for taking an exceptionally strict line with employees.

Michael Nuñez, the Gizmodo writer, did not have a long history of reporting on Facebook and had never lived in the San Francisco Bay Area. But the twenty-eight-year-old journalist had a connec-tion at the company. On the morning of February 25, he was sit-ting at his desk at Gizmodo's Manhattan offices when he saw a message pop up on Google's messaging service, Gchat, from his former roommate Ben Fearnow. The two were in daily contact over Gchat. Fearnow, who had recently become a contract em-ployee for Facebook in its New York offices, enjoyed goading and teasing his former roommate. Fearnow was part of a small group of former journalism students and reporters whom Facebook had hired to try to burnish its news sense, but as a contract worker, he was not fully part of the company culture at MPK; nor did he receive the benefits of a full-time Facebook employee. "I'm assuming you've seen this?" he wrote to Nuñez and attached what looked like a Facebook post written by Zuckerberg.

It was a memo to all employees that had been posted on the general Workplace group. In the post, Zuckerberg addressed the brewing scandal at the company's Menlo Park headquarters over the defacement of Black Lives Matter slogans written on the walls of its office buildings. "'Black Lives Matter' doesn't mean other lives don't," he wrote, pointing out that the company had an open policy of what employees could write on the office walls or put up in poster form. But "crossing out something means silencing speech, or that one person's speech is more important than another's."

Nuñez was initially conflicted about publishing the memo. Was it news that Facebook was experiencing the same type of racial tensions that were being seen across the United States? He asked some editors, and they decided that readers would be intrigued by insights into Facebook's internal culture.

The story, published on February 25, at 12:42 p.m., did not look good for Zuckerberg. His own words were being used to go after his employees, painting a picture of Facebook employees as privileged elitists out of touch with the Black Lives Matter movement. The call went down to Ahuja, the so-called rat catcher, to find the Facebook employee responsible for the leak.

Ahuja's team searched email and phone records of Facebook employees to see who might have been in contact with Nuñez. The group could easily view messages written on Facebook, but few employees were likely to have been naïve enough to message a member of the press from their own Facebook pages. They might, however, have used WhatsApp. Those messages were end-to-end encrypted, so the investigations team could not see the contents of messages, but they could retrieve data, such as which two numbers had messaged each other and when the correspondence had taken place.

The team had other tools as well. Through the location permissions on phone apps, Ahuja and her team had a fairly accurate record of where Facebook employees traveled. They could check, for instance, if the phone of a Facebook employee had been in close proximity to the phone of a reporter from a news outlet that had recently run a scoop about the company. Facebook also had the technical means to track anything an employee typed or hovered over with their mouse on their work laptop. It could create a detailed map showing exactly what internal documents or memos Facebook employees had spent time reading, and if anything appeared unusual—if an employee from a department was spending an unusual amount of time reading up on memos from another department, for example—it could flag the irregularity.

Once Ahuja zeroed in on her suspect, the employee was called in for what he or she was told would be a general meeting pertaining to an investigation. Ahuja would be waiting there to grill them. She would often remind the employees of the stock they might have forfeited or the NDAs they might have violated by leaking information about the company. "She was very impersonal, to the fact. She took her job seriously, and if you broke Facebook's rules and leaked anything about the company, she had no mercy," said one former employee who left the company after admitting to leaking material to a journalist.

On March 4, eight days after Nuñez's story on Black Lives Matter was published, Fearnow awoke to a series of messages on his phone asking him to come into the office for a meeting. It was his day off, but he wrote back and offered to join a virtual call via his laptop.

Soon, Fearnow was on a conference line with Ahuja. She immediately got to the point. Had he been the source of the leaks to Gizmodo? Initially, he denied the accusation, but Ahuja began reading him a series of messages between him and Nuñez that she claimed had taken place on Facebook Messenger. Fearnow was confused; he knew he would not have been so careless as to use Facebook's own messaging system to speak to Nuñez.

"I picked up my phone and looked through Gchat and realized that she was reading me my Gmail conversation with Nuñez," Fearnow recalled. "I had no idea they would be able to see that. I was going to deny the accusation, but in that moment, I knew I was screwed."

Fearnow asked Ahuja how she had accessed his Gchat. She told him she could "neither confirm nor deny" where she had seen the messages, a response that made him laugh, given that he was looking at the messages within Gchat as they spoke. He realized he was being fired and had no chance to question her further.

Fearnow was not the only person Ahuja called that morning. He and Nuñez had shared an apartment with another roommate, Ryan Villarreal, in Brooklyn. Villarreal had also gone on to work at Facebook with Fearnow. He knew nothing about the leaks or Fearnow's connection to them. He had simply seen Nuñez post the Gizmodo article he had written to Facebook and had "liked" it, wondering to himself who the source had been.

Hours after Fearnow was fired, Ahuja emailed Villarreal to set up a meeting. He was shown Nuñez's Facebook page and asked why he had liked the story Nuñez had written about Zuckerberg's Black Lives Matter memo. He struggled to respond and eventually answered that he didn't even remember liking the post. But he, too, was told he no longer had a job as a contractor at the company.

Nuñez felt bad for his friends. But by then, dozens of other Facebook employees were reaching out to him. Some only wanted to confirm stories Nuñez had already published, while others told him he only knew the "tip of the iceberg."

Employee anger with Zuckerberg was steadily rising as Trump's presidential ambitions shifted from a fantasy to a credible possibility. On March 3, a Facebook worker decided it was time to take the matter up with Zuckerberg directly. He entered a question for the next all-hands meeting. "What responsibility does Facebook have to help prevent a President Trump in 2017?" By mid-morning the next day, his question had received enough votes from employees to make it to the short list of questions for Zuckerberg.

Facebook's corporate communications team panicked. If the question were made public, Facebook would live up to its reputation as a company of liberal Democrats. It would look like the platform was biased against Trump. Whatever answer Zuckerberg gave would be highly controversial, and newsworthy. Some of the PR staff tried to sway the all-hands votes by voting for other questions. But by midday, the question was still at the top of the list. They had to prepare Zuckerberg.

They advised him to avoid answering the question. He should focus instead on Facebook's commitment to free speech and on the role the company played in a democratic society. He should outline Facebook's strategy around the election and the way it was committed to supporting any presidential candidate. Most important, he should avoid giving any answer that implied that Facebook preferred one candidate over another. It made good business sense. Facebook was not a publisher, with employees making editorial decisions. It was a technology company that simply hosted the ideas that were posted by its users. That refrain, which all social media companies fell back upon, protected it from defamation suits and other legal liabilities and kept the company out of the fray of partisan politics.

That afternoon, Zuckerberg redirected his answer, again, to his belief in free expression. The story got out anyway. On April 15, 2016, Gizmodo published an article by Nuñez headlined "Facebook

Employees Asked Mark Zuckerberg If They Should Try to Stop a Donald Trump Presidency." The story included a screenshot of the question and focused on Facebook's power as a gatekeeper of information. "With Facebook, we don't know what we're not seeing. We don't know what the bias is or how that might be affecting how we see the world," Nuñez wrote.

The story was widely shared by conservatives and became fodder on Fox News and right-wing blogs that had been accusing Facebook of censoring Trump and his supporters. But more alarming to Zuckerberg was the fact that an account of the private all-hands meeting had been leaked to the media. This had never happened before. In Facebook's eleven years of operation, internal conversations, especially Zuckerberg's all-hands meetings, had been considered sacrosanct.

Nuñez already had his sights set on another line of reporting, one that would reveal the inner workings of a mysterious new section Facebook had introduced to help it compete against newcomers in the social media space. When people wanted news, they were more likely to turn to sites like Twitter or YouTube. Both offered anyone the ability to create and view content, and both seemed popular with newsmakers. Increasingly, top trending topics and videos on those sites seemed to drive daily news events. The more highly they featured on the day's trending news, the more news reports cited them, driving even more people to search and view the topics.

Zuckerberg was worried. He wanted people coming to Facebook for everything, including the news. "When something big happened in the world, research showed us that people were more likely to go to Google or Twitter. We wanted to be players in the news division, too," recalled one Facebook executive.

Some executives saw the answer in a feature called "Trending

Topics." It was the first thing U.S. users saw on the top right-hand side of the page when they logged into Facebook. Beneath an inconspicuous black headline reading "Trending," squiggly blue arrows highlighted three key topics that Facebook users were sharing information on that day. Engineers who designed the feature said the topics were a reflection of what their algorithms had surfaced as the most popular issues of the day.

But there was confusion over what metric Facebook used to determine the Trending Topics. It may have claimed that the system was governed by an algorithm, but the people hired to work on the section all appeared to have backgrounds in journalism. At times, the topics highlighted seemed too esoteric, or convenient. While it would seem nice that on Earth Day users were discussing clean water and recycling, some Facebook employees questioned whether those topics were organically trending or were being manipulated by Facebook.

For months, Nuñez had been circling the employees hired by Facebook to run its Trending Topics section. Beyond his two former roommates, Fearnow and Villarreal, he had met a handful of other recent college graduates hired to join the team. Using Google, he had created a document that listed all the employees on the Trending Topics team and divided them by those who were "drinking the Facebook Kool-Aid" and those he could approach as a possible source. From those who agreed to speak, he was hearing a disconcertingly similar story. "They would talk about how disturbing it was that they were making these decisions about what people saw in Trending and what they didn't," Nuñez said. "Facebook made it seem like it was a black box, like there were systems making these decisions, but actually there were employees who were making critical decisions without any oversight."

On May 3, 2016, Nuñez published the first in-depth look at the Trending Topics team. The story gave an inside look at what

Nuñez called "grueling work conditions, humiliating treatment, and a secretive, imperious culture." He described how a small group of contract employees in New York, many of them recruited out of journalism school for their "news experience," were given the role of "curators" and strictly instructed not to describe their jobs as editorial.

The system was set up to give the Trending Topics team a list of topics that users were discussing in real time. But when curators logged in for their eight-hour shifts, they found that much of what was surfaced had to do with the latest dress a Kardashian was wearing or a contestant on *The Bachelor*. It was not the type of debate Facebook had envisioned featuring on its platform.

Curators were instructed to use their individual judgment when selecting topics to promote. After creating a label and a description for each topic, they were to assign it a value that would make it more or less likely to appear on a user's page. They were to "blacklist" certain topics, such as those that appeared too frequently or were considered irrelevant, like the aforementioned Kardashians and *The Bachelor*. They were also told to remove topics that were duplicates or that did not reflect the current news cycle.

Initially, Nuñez was told, there was little oversight. "The people who were making these decisions were largely young and straight out of college. It seemed like too big of a responsibility," noted one former member of the Trending Topics team. Many on the team were disturbed by the power they were given to make decisions over what hundreds of millions of Facebook users saw. After Nuñez's initial story went viral, these team members became even more introspective about the work they were doing. Several weeks after Nuñez had published, a member of the team reached out to him in an anonymous email. After exchanging messages proving they worked at Facebook, the source agreed to speak with Nuñez by phone. Over the course of a weekend, the source and Nuñez communicated nonstop, as the source explained to Nuñez how the system worked and how easily curators could shape the topics that were trending on Facebook.

"One example was Paul Ryan. They [Facebook engineers] would straight up say Paul Ryan is trending too much," said Nuñez, referring to the former Speaker of the House of Representatives. "So, they would keep tweaking the tool until Paul Ryan would stop showing up."

The source that came forward to speak to Nuñez identified as politically conservative. He believed the public had a right to know about the decisions being made by a small group of people at Facebook. And the article published as a result of those conversations, on the morning of Monday, May 9, 2016, delivered. Titled "Former Facebook Workers: We Routinely Suppressed Conservative News," the bombshell story appeared to confirm suspicions long held by conservatives and long voiced by far-right political pundits like Glenn Beck and Trump campaign leader Steve Bannon. It claimed that Facebook and other gatekeepers of information on the internet were putting their thumbs on the scale.

The piece had a controversial reception, with some questioning its agenda. And within Facebook, members of the Trending Topics team were furious at how the article portrayed their work. While it was true that some of them were uncomfortable with the power they had been given, they were not pushing a liberal or conservative agenda. Their mandate had been to ensure the Trending Topics section did not appear overly repetitive. Now their decisions were being painted as political.

The decision earlier that year to leave Trump's account untouched had avoided a confrontation with conservatives. But now Facebook was in the hot seat. "It is beyond disturbing to learn that this power is being used to silence view points and stories that don't fit someone else's agenda," the Republican National Committee pronounced in a statement.

At MPK, Sandberg and Zuckerberg directed their security team to find out who else was leaking to Nuñez, but staff turnover at Trending Topics was high, and most of the people on the team were contractors with little loyalty to the Facebook brand. Zuckerberg turned to his engineers and asked them to review the way Trending Topics worked. Was there a way to ensure, ultimately, that only algorithms decided which topics Facebook users saw?

Sandberg, meanwhile, focused on the PR crisis. Something had to be done to appease the right, but she wasn't the ideal messenger. Her ties to the Democrats made her a liability; she was rumored to be on a short list for a cabinet position in the likely eventuality of a Clinton administration. She was also focused on writing *Option B*, the follow-up to her 2013 best seller, *Lean In*, which focused on empowering women in the workplace. The new book, a highly personal guide to building resilience from loss, came from a very different place.

In May 2015, her husband, Dave Goldberg, had succumbed to cardiac arrhythmia while the couple was in Mexico for a friend's fiftieth birthday celebration. The loss was devastating for Sandberg. Zuckerberg proved to be a source of great support; he helped arrange Goldberg's memorial service and encouraged Sandberg in moments of doubt about her ability to lead again as she returned to work. She had begun to write *Option B* as a way of processing her grief while still overseeing the day-to-day operations of the company. In it, she recounted her rocky transition back to work and Zuckerberg's full support following her loss. The two executives had never been closer, but to Sandberg's team, she appeared distracted and short-tempered at times. Employees were walking on eggshells, wary of upsetting her.

In June 2016, there had been a particularly unpleasant incident that highlighted the pressure Sandberg was under as she dealt with her loss. She was in Washington, DC, on one of her quarterly visits to rub elbows with government officials and the press. Facebook's lobbyists had set up a room in the Capitol Building as a pop-up for the company to showcase demonstrations of its new products, including the Oculus VR headset. The room was reserved for two cocktail hours, the first for members of Congress and the second for journalists. Sandberg was the guest of honor at both. She was coming off two days of back-to-back meetings with lawmakers and a speech at the conservative American Enterprise Institute to try to defuse concerns of bias raised from the Trending Topics controversy. The cocktail hours were her last obligations before she could return home.

When she arrived at the Capitol, she was annoyed when she learned she wouldn't have a chance to meet with Facebook's employees from the Washington office before the event began. Also, she had asked to take a quick rest somewhere quiet, but it emerged that the lobbyists and Sandberg's chief aide had neglected to prepare a private greenroom for her between the events.

"How could you forget? I can't believe it!" she yelled at the aide in the hallway outside the room, in full view of several Washington-based employees. She then turned to a new employee, who was in training to replace her chief aide, and shouted that it was a gross oversight that couldn't be repeated. "It was a full dressing-down," one Washington-based employee recalled. Minutes later, the two young women could be heard crying in the bathroom. Greg Mauer, who had worked for former House Speaker John Boehner, asked his former colleagues to find a room in the building for Sandberg. The problem was solved, but the episode sent a sobering message.

The COO had become easily angered and impatient after Goldberg's death; staff were afraid of upsetting her. It had become a management problem, one former employee recalled: "No one wanted to speak up after that."

When it came to the sticky question of how to handle their conservative critics, Sandberg turned, as was often the case, to Joel Kaplan for advice. He assured her that he could handle the controversy by convening a meeting of top conservative media executives, think tank leaders, and pundits in Menlo Park. It was critical, however, that Zuckerberg agree to play a prominent role in the event.

On May 18, sixteen prominent conservative media personalities and thought leaders—including Glenn Beck of Blaze TV, Arthur Brooks of the American Enterprise Institute, and Jenny Beth Martin of Tea Party Patriots—flew to Menlo Park for the afternoon gathering to air their concerns of political bias. Only Republican Facebook staff were allowed in the room, a former Facebook official said. That meant Peter Thiel, the company's vocal pro-Trump board member, and a smattering of Washington staff, including Joel Kaplan. Zuckerberg, who over the years had made a point of never publicly declaring his political affiliation, led the ninety-minute meeting over cheese, fruit, and coffee. The guests politely aired their grievances. Some attendees urged Zuckerberg to hire more conservatives. Brooks of AEI warned that the platform should remain neutral on political content; Zuckerberg shouldn't allow it to become a "monoculture" of a particular political or religious bent.

Zuckerberg was effusive in his embrace of all political speech and promised that the company did not intend to stifle conservative views. He assured his guests that giving voice to all political viewpoints was good for the mission and the business.

Having an audience with the company's CEO appeared to help. After the meeting, the group went on a tour of the campus and sat through a demonstration of the Oculus headset. Beck later wrote on his blog that Zuckerberg "really impressed me with his manner, his ability to manage the room, his thoughtfulness, his directness, and what seemed to be his earnest desire to 'connect the world.'"

Beck and the other attendees left MPK somewhat mollified; some said they believed Mr. Zuckerberg would make an earnest effort to make the platform neutral. But the event was deeply divisive inside the company. Members of the policy, communications, and engineering staff feared the gathering would open the door for more interest groups to press their agenda with Facebook. They complained to top leaders that the event was shortsighted and wouldn't fully extinguish suspicions of political bias.

Worst of all, they believed the meeting had legitimized the people responsible for propagating damaging conspiracy theories. Beck had falsely accused a Saudi national of involvement in the Boston Marathon bombing; he alleged that the Obama administration needed another Oklahoma City type of attack to improve its political standing. Beck's theories were amplified further by right-wing figures like Alex Jones, who had amassed hundreds of thousands of followers on Facebook. One of Jones's most popular and controversial positions was that the December 2012 massacre at Sandy Hook Elementary School, in Newtown, Connecticut, had been faked and that families who had lost children in the shooting were paid actors. Those followers were directed to Jones's website and radio show, where he made millions by selling branded merchandise.

"We weren't giving Black Lives Matter an audience with Mark, but we were giving that access to conservatives like Glenn Beck? It was such a bad decision," a former employee said.

The meeting was a turning point for Facebook, which until then had pledged political neutrality. And it marked a pivotal moment for Sandberg, whose decisions would become increasingly dictated by fears that a conservative backlash would harm the company's reputation and invite greater government scrutiny.

"At that point in the campaign, no one expected Trump to win. But Republicans controlled the House and Senate and could make life difficult for tech companies on the hill," observed Nu Wexler, a former DC-based spokesman for Facebook. "So, rather than denying their bad faith accusations and producing hard data that refuted them, Facebook engaged their critics and validated them. Going forward, this became the modus operandi for all bias accusations from the right."

As the 2016 presidential campaigns kicked into high gear, Facebook's News Feed showed that Americans were more divided than ever. The race between Hillary Clinton and Donald Trump was a feeding frenzy for highly partisan news, with each side eager to demonize the other. But an even more troubling trend was a number of obviously false news sites that were pumping out increasingly outlandish stories about the two candidates. Throughout the United States, and in countries as far flung as Macedonia, enterprising young people realized they could make money feeding Americans exactly the type of content they were craving. Suddenly, stories claiming that Hillary Clinton was secretly in a coma, or that Bill Clinton had fathered a child out of wedlock, were running rampant across Facebook. The people behind them were largely apolitical, but they knew that the more outlandish the story, the more likely it was that a user would click on the link.

Employees who worked on the News Feed team raised the issue with managers, only to be told that false news didn't run afoul of Facebook rules. This didn't sit well with them. "I was grinding my teeth over that; we were seeing all these junk sites sitting really prominently in people's feed. We knew people were opening

Facebook and seeing totally fake news stories at the top of their home page, but we kept being told that there was nothing we could do—people could share whatever they wanted to share," recalled one former News Feed employee. At a meeting with his manager in early June, he pointed out that many of the sites promoting false or misleading articles on Facebook appeared to be employing dubious methods to promote their stories. He suspected they were using fake Facebook accounts. "I was told it was being looked into, but it wasn't. No one ever followed up with me."

In fact, Chris Cox and other Facebook engineers attempted to address the problem that same month by introducing yet another algorithm change, one in which the content of family and friends was weighted above all else. But outside researchers found that this change reaped unintended consequences. By prioritizing family and friends, Facebook was deprioritizing accredited news sites like CNN and the *Washington Post.* Users no longer saw posts from news sites featured prominently on their News Feeds, but they continued to see the false and hyper-partisan news that their family members and friends were posting.

In the company's weekly Q&A, the topic of the News Feed was raised again and again by employees concerned about the kinds of stories going viral on the platform. The topic came up so frequently that on June 18, 2016, Boz posted a memo on one of the Workplace groups addressing the subject of Facebook's responsibility to its users.

"We connect people. Period. That's why all the work we do in growth is justified. All the questionable contact importing practices. All the subtle language that helps people stay searchable by friends. All of the work we do to bring more communication in. The work we will likely have to do in China someday. All of it," Boz wrote.

"So we connect more people," he declaimed in another section of the memo. "That can be bad if they make it negative. Maybe it costs someone a life by exposing someone to bullies. Maybe someone dies in a terrorist attack coordinated on our tools. And still we connect people. The ugly truth is that we believe in connecting people so deeply that anything that allows us to connect more people more often is *de facto* good."

The title of the memo was "The Ugly."

The debate within the Workplace groups was fierce. Some employees defended the memo, arguing that Boz was just voicing an uncomfortable truth and that Facebook was a for-profit company that had to prioritize business first. Most, however, expressed discomfort with the position taken in the memo and voiced concern that it exposed the cold calculus being made by the company's top brass.

As Facebook employees surveyed what appeared to be a global rise in hate speech, they found the name of their company surfacing repeatedly as a source of conspiracies, false news accounts, and organized campaigns of hate speech against minorities. Trump's Muslim ban announcement was used by far-right leaders across the world to take more extreme positions on Muslim immigrants and refugees. In Myanmar, several of the country's military figures pointed to Trump's Facebook posting through their own Facebook pages and argued that if the United States could ban Muslims, Myanmar should do the same. And indeed, human rights activists increasingly linked the platform to attacks against Myanmar's stateless Rohingya minority and to the brutal crackdown on civilians by the recently elected president of the Philippines, Rodrigo Duterte.

That next all-hands was packed. That week, there was no doubt as to the line of questioning.

Boz had already defended his post on Facebook's Workplace, arguing that he didn't actually agree with aspects of his own memo and had written it to inspire debate. But employees wanted more. They wanted to know if he had actually considered the lives lost in countries where Facebook had grown at an astronomical pace. They asked if Facebook users from those countries had responded to his post directly, and if he felt bad about what he had written. Boz looked remorseful, but he repeated that he was simply making an intellectual argument and had intended the memo to spur debate.

Zuckerberg and Sandberg assured employees that they never wanted to see Facebook used to do harm in the world and that they were not devoted to "growth at all costs." "I want to only see us commit to responsible growth," one employee recalled Sandberg saying. He said he left feeling skeptical. "She said all the right things, but I don't think most people were convinced," he recalled. "I wasn't convinced."

Over the past four years, the partnership between Zuckerberg and Sandberg had found its natural rhythm. When, in June 2012, Sandberg was appointed to Facebook's board of directors, the first female member to join Marc Andreessen, Erskine Bowles, Jim Breyer, Don Graham, Reed Hastings, and Peter Thiel, Zuckerberg gushed in the press release, "Sheryl has been my partner in running Facebook and has been central to our growth and success over the years." Sandberg, in turn, confirmed her dedication to the company. "Facebook is working every day to make the world more open and connected. It's a mission that I'm deeply passionate about, and I feel fortunate to be part of a company that is having such a profound impact in the world."

They continued to meet regularly, Mondays and Fridays, by phone, if needed, as they both traveled more frequently. Their old squabbles over resources dissipated. As Marc Andreessen described it, theirs was the perfect pairing: the founder who could stay focused on the big vision and the partner who could execute the business plan. "Her name has become a job title. Every company we work with wants 'a Sheryl,'" Andreessen said in an interview with *Fortune* magazine. "I keep explaining to people that we haven't yet figured out how to clone her."

Sandberg had developed a new business of data mining at scale, and the money was rolling in. She was cultivating new ad tools, including one called "Custom Audiences," which would allow brands and political campaigns to merge email lists and other data with Facebook data for richer targeting. The company was pursuing an automated auction that would process millions of bids every second for ads based on Facebook data and a user's browsing history while off the site. It was also working on a tool called "Lookalike Audiences," in which brands would provide lists of their customers and Facebook would identify users with similar profiles for ad targeting.

While some thought Sandberg had adapted to fit into the Facebook culture—"Google was a values-driven organization. Facebook was absolutely not a culture driven by values," a former Google employee who worked with Sandberg said—others believed she had changed the culture of the company. This was perhaps most evident in the people she hired. Kaplan was one example. Elliot Schrage was another.

Schrage had been one of her first hires. He stood out in Facebook's freewheeling youth culture. He was twenty-four years older than Zuckerberg, and his idea of casual dress was unfastening the top button of his crisp dress shirts. He was an intimidating figure; with his piercing eyes; rectangular glasses; dark, curly hair; and low, gravelly voice, he bore a striking resemblance to film director Sydney Pollack.

Sandberg and Schrage, a Chicago native with a Harvard Law degree and a wry sense of humor, had bonded at Google. They were non-techies in an industry that looked down on MBAs and lawyers. For them, Davos, not *TechCrunch* Disrupt or the *Wall Street Journal*'s D conference, was the highlight of their year. And both voraciously kept up with the who's-up-who's-down rhythm of national politics.

In public, Schrage presented a friendly and earnest demeanor, though he was tough with reporters and known to quibble over details in stories, sometimes barking on the phone with editors about word choices in headlines. In May 2010, a glitch in Facebook's systems allowed some users to peer into the private chat messages of other users. Schrage agreed to take questions from readers in *Bits*, a blog for the *New York Times*. He fielded roughly three hundred angry questions about the growing number of data abuses. "Reading the questions was a painful but productive exercise," he said. "Part of that pain comes from empathy. Nobody at Facebook wants to make our users' lives more difficult. We want to make our users' lives better."

But behind the scenes, as witnessed by his performance in Senator Schumer's office in 2009, Schrage executed a different part of the company's public relations strategy. He employed more aggressive tactics to fight off the threat of regulations, some members of the communications and policy staff said. He saw it differently: "I felt the team needed to move beyond its historic roots in product communications and become more professional, better equipped to understand and address the privacy, regulatory, competitive, financial and other policy issues confronting Facebook," Schrage said.

He also operated as a cudgel against critics who were complaining not only about what Facebook took away from users (data) but also about what users put *onto* the platform. As the platform grew, so, too, did the level of toxic content, which seemed particularly out of control among younger users.

In late January 2012, Bill Price, chairman of the board for the nonprofit child advocacy group Common Sense Media, described a meeting with Sandberg at Facebook's new offices at One Hacker Way, in Menlo Park. The cofounder of San Francisco–based TPG Capital, one of the largest private equity companies in the nation, Price moved in the same circles of wealthy and politically connected Bay Area business leaders as Sandberg, serving alongside tech executives on the boards of organizations like the California Academy of Sciences.

But Common Sense Media was a real public relations problem for Facebook. The organization was a vociferous critic and part of a rising tide of protest from consumer, child advocacy, and privacy protection groups that were scrutinizing how Facebook was affecting young people and what it was doing with private data. Founder Jim Steyer, a former civil rights attorney and professor at Stanford, blasted the company in news articles and had just completed a book called *Talking Back to Facebook*, a critique of social media with regard to the social and emotional well-being of youth.

Since opening the platform to high school students in 2005, Facebook had faced concerns about youth safety and mental health. When not compulsively checking their posts for comments and likes, young users competed over who had the most friends. Reports of cyberbullying multiplied. Facebook ignored its own rule prohibiting children under the age of thirteen from holding accounts. In May 2011, *Consumer Reports* estimated that 7.5 million underage users were on the platform.

As a condition for the meeting, according to Price, Sandberg had insisted that Steyer not participate. Price agreed, but not without issuing his own veto: Schrage. Yet, when Sandberg greeted Price with an embrace and a warm smile, he saw that Schrage was right behind her. Firmly, but without raising his voice, Price looked at Sandberg and asked, "What's Elliot doing here?"

Schrage was in charge of policy, Sandberg reminded Price, and therefore the best person to speak for Facebook on youth-related issues. Price was visibly irritated, but he moved on. He had an important message to deliver and wanted their help, he told Sandberg. After he sat down, Sandberg slid into a leather chair next to Price, kicked off her shoes, and folded her legs under her. On the other side of the conference table, Schrage sat alone and played bad cop, folding his arms across his chest and glaring.

Price started the meeting by saying he wanted a reset. Surely, Common Sense and Facebook could agree on some way to protect children online. The nonprofit was concerned in particular about kids posting content that would come back to haunt them: the tipsy teen holding a beer can at a party or the young girl posing for suggestive selfies. He raised the idea of an "Eraser button." Why not create a tool for youth to give themselves a second chance? It seemed like a reasonable idea. Sandberg's kids weren't old enough for her to relate. But if Schrage, the father of children old enough to be on Facebook, should have been the most persuadable, he quickly put that idea to rest.

"It's not possible," he huffed, according to Price. "You don't understand how the technology works."

Price was put off by Schrage's tone, but he wasn't intimidated. He pushed back. Why wasn't it possible? He wanted to understand why Facebook, with top engineers and billions of dollars in revenue, couldn't develop the fix.

Sandberg said nothing.

An Eraser button would have harmful ripple effects across the internet, Schrage said. "You don't understand how the internet works!"

When Sandberg finally spoke, she tried to put Price on the defensive, claiming that Common Sense was unfairly targeting Facebook. "Google has just as many problems, if not more," she said.

And then, wearing a hurt expression, she changed her tone and brought up Steyer's book, Price recalled. Why did he title it *Talking Back to Facebook*? she asked. Google and other companies had the same data collection practices as Facebook, didn't they? Wasn't the book just a self-promotion vehicle for Steyer?

Google was partnering with Common Sense on safety and privacy programs, Price explained. It hardly mattered, he realized at that point. He had come to Facebook to seek a partnership. The Facebook executives were sophisticated and smart; they surely understood Common Sense's concerns. But clearly, they had different agendas. Sandberg was there to complain about Steyer's book title. Schrage was there to attack someone he viewed as an opponent. "You'll be on the wrong side of history if you don't try harder to protect kids," Price said, switching tack.

At that, Schrage jumped in with his own warning. Price didn't want to make an enemy of the local Goliath. Many of Common Sense's board members were tech investors, and they wouldn't look favorably on the prospect of losing Facebook as a business partner. Facebook was gearing up for an IPO, which the Facebook executives were preparing to announce the next day. The subtext was clear: "I felt that they were attempting to intimidate and threaten that if we continued to pursue these issues, they would act in a way that was aggressive toward Common Sense and our board," Price recalled.

Sandberg was silent as Schrage continued to level accusations at the visitor. When the meeting broke up, she gave Price another perfunctory embrace. Price and Schrage didn't bother shaking hands.

Schrage later denied threatening Price. "It's laughable to suggest that I would be in a position to threaten or intimidate Bill Price, a founder of one of the largest and most successful private equity funds in the world. Why would I do that? What is my upside? Facebook's?" Schrage said. But Price had a very different takeaway:

"I left the meeting feeling pretty depressed," he recalled. "It's sort of a typical story of where ultimate power corrupts people's perspective."

Sandberg's dominion at Facebook had steadily expanded. She had recently gotten more involved in an important part of the business: Facebook's content moderation strategies.

The team originally sat under the operations department. When Dave Willner joined in 2008, it was a small group of anywhere between six and twelve people, who added to the running list of content the platform had banned. The list was straightforward (no nudity, no terrorism), but it included no explanation for how the team came to their decisions. "If you ran into something weird you hadn't seen before, the list didn't tell you what to do. There was policy but no underlying theology," said Willner. "One of our objectives was: Don't be a radio for a future Hitler."

Zuckerberg rarely involved himself in the content moderation team's decision making or policies, or seemed aware of the arguments shaping the rules for what millions of Facebook users were and were not allowed to post on the platform. "He mostly ignored the team, unless there was a problem or something made news," recalled one early member. "I would not say he was in there arguing about what the First Amendment meant, or where Facebook had a role in deciding free-speech policies for the rest of the world. That was not something he cared deeply about."

The team, however, was increasingly landing Facebook in hot water. As early as 2008, the company had been battling with moms across the world over their right to share photos showing themselves breastfeeding their babies. The decision over banning a pro-democracy group in Hong Kong in 2010 was met with front-page headlines and protests. The final straw came in late 2011, when Willner and the rest of his team, which had enlarged significantly, made a decision not to remove a Facebook page with a name featuring a crass joke about the daughters of Tony Abbott, the Australian Opposition leader.

Left to deal with a fuming Abbott, Sandberg began to assert herself more in the content moderation policy under her purview. The decision took many on the team by surprise. "She was saying that she, a person who has a political calculus over Facebook's lobbying arm, wanted to oversee decisions," said one team member. "It didn't take a genius to know that the politicians and presidents who called her complaining about x or y decision would suddenly have a direct line into how those decisions were made."

Willner left the company shortly afterward. "We should have seen the writing on the wall," said the team member. "Facebook was a major player on the world stage, and I guess it was overly idealistic that we could keep that political calculus out of our content decisions."

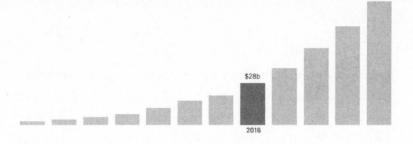

$28b

2016

Chapter 5

The Warrant Canary

Ned Moran was staring at his laptop watching a conversation unfold between a Russian hacker and an American journalist. A security analyst who worked with a specialized group at Facebook known as the threat intelligence team, Moran had earned a name among cybersecurity professionals for his prodigious knowledge of and experience in studying aggressive, state-backed hackers. Tall, bespectacled, and bearded, he was a man of few words; when he did speak, his words were so softly and deliberately spoken that people stopped what they were doing to lean in and listen. His reputation guaranteed that whatever he was about to say was worth hearing.

Moran knew more about foreign-backed hackers than nearly any other cybersecurity professional, but even he had not seen an exchange between hackers and a journalist target before. At times, minutes passed between messages as he waited, along with the American journalist, to see what the Russian would say next. From his perch in Facebook's DC office, he knew the identities and locations of the two users trading messages. Ever since he discovered the DCLeaks page earlier that August, he had been obsessively reading all its chats. In any other circumstance, he would not have been spying on the real-time communications of a journalist. But he had been following the Russians across Facebook and had seen when they started up a chat in Facebook Messenger with the journalist. Moran saw what type of devices they were using and what type of searches they were running within Facebook. He knew that the Facebook page calling itself "DCLeaks" was a Russian asset. Just weeks before, he had uncovered the page while following a trail of bread crumbs left by the Russians as they sought to establish Facebook pages, groups, and accounts before the U.S. elections. The Russians had created DCLeaks on June 8, and now they were using it to try to lure a U.S. journalist into publishing documents that the same group of Russian hackers had stolen from the Democratic Party.

It was less than three months before the United States was set to vote in an increasingly heated presidential election between Donald Trump and Hillary Clinton, and Moran had hard evidence that the Russians were doing exactly what some U.S. intelligence officials suspected: hacking the Clinton campaign and then releasing critical emails to try to embarrass the Democratic front-runner for president. It was an unprecedented moment of espionage, breaking every norm of cyberwarfare previously established. Moran knew it was significant, and he reported it to his superiors.

Just a few miles away from where Moran sat, U.S. intelligence officers were scrambling to learn as much as possible about the Russian hackers who had infiltrated the Clinton campaign. As experienced as the intelligence officers were, they lacked Facebook's bird's-eye view. For Moran and the rest of Facebook's threat intelligence team, part of the draw of the job was the overview the platform afforded them; many had previously worked at the NSA, the FBI, and other branches of government, studying the very hackers they now had under surveillance.

Moran had been paying closer attention to the Russians since early March, when he and another Facebook security analyst discovered that they were attempting to hack Facebook accounts across the United States. Independently, the U.S. accounts had little in common. But when they were viewed together, a pattern emerged: they were all connected to people tied to but not directly involved with the 2016 presidential election. These included family members of candidates and political lobbyists—the wife of a major Democratic lobbyist and the children of GOP presidential candidates were among those targeted.

Facebook's security team had anticipated that the Russians would ramp up surveillance of presidential candidates ahead of the 2016 elections. Yet no one at the top appeared to grasp the severity of the Russian activity. Moran was sending weekly logs and reports to Alex Stamos and filing them on Facebook Workplace groups. Stamos was sharing them with his boss, Colin Stretch, Facebook's general counsel, who, he understood, would pass on the highlights to Elliot Schrage and Joel Kaplan, two of the most powerful men at Facebook. It was at their discretion that Zuckerberg and Sandberg were briefed on problems as they arose.

Over the years, the security team had largely been sidelined by the two leaders, who didn't take an active interest in their work or solicit their reports. "They were briefed on major issues, like when hackers from North Korea or China tried to infiltrate the platform," said one security team engineer. "But it was not like they asked for regular meetings on security threats, or even sought out the opinion of the team. They honestly treated security like it was something they wanted taken care of quietly, in a corner, where they didn't have to regularly think about it."

Moran and other members of the threat intelligence team were in DC, while the rest of the team was relegated to a small building on the outskirts of MPK's sprawling campus. Regardless of how many dotted lines led to Sandberg, the company was still at its heart focused on engineers. Teams that developed new products, pushed for growth, and tracked the number of hours Facebook users spent on the platform each day reported directly to Zuckerberg. They were busy that spring carrying out his audacious ten-year vision for the company, which included major moves in the fields of artificial intelligence and virtual reality.

The rest of the world was slowly discovering the scope of the Russian activity. On June 14, CrowdStrike, a cybersecurity firm hired by the Democratic National Committee, announced that it had found conclusive evidence that hackers tied to the Russian government had hacked and obtained emails from senior members of the Democratic Party, including people in the Clinton campaign. Within weeks, two more cybersecurity companies, ThreatConnect and Secureworks, joined CrowdStrike in publishing independent reports with technical details on the attackers' methods. The reports drew a picture of how the Russians had tricked the Democratic campaign staffers into revealing their email passwords. Separately, suspicious accounts with names like Guccifer 2.0 had suddenly appeared on Twitter and other social media platforms, offering to share the hacked documents with interested journalists.

The reports synched with what Moran had been seeing. The sheer scope of the Russian hackers' plan was like nothing that had ever been carried out before on U.S. soil. These were the same hackers he had been watching for years, and while many of the techniques were familiar, others were not. His team knew the Russians had been probing accounts of people close to the Clinton campaign; now Moran wanted to go back and find out if they had missed anything on the platform. The reports gave the security team a guide to search for the Russians within Facebook's systems.

Using details from the reports, Moran had identified the suspicious page under the name "DCLeaks." Since its creation, the page had shared stories about the DNC hack. His team found similar accounts, including one that had created a Facebook page called "Fancy Bear Hack Team." CrowdStrike and the other cybersecurity companies had dubbed one team of Russian government hackers "Fancy Bear," and the English-speaking press had adopted the catchy name—the Russian hackers using the name on their own Facebook page seemed to be mocking Facebook's inability to find them. The page contained data stolen from the World Anti-Doping Agency, another arrow that pointed to the Kremlin; in past years, Russia had been caught doping its athletes and using other ploys to try to win medals at the Olympics and other sports competitions.

Facebook had no playbook for the Russian hackers, no policies for what to do if a rogue account spread stolen emails across the platform to influence U.S. news coverage. The evidence was clear: Russian hackers posing as Americans were setting up Facebook groups and coordinating with one another to manipulate U.S. citizens. But Facebook didn't have a rule against it.

Ironically, the hackers were using Facebook exactly as it was intended. They were making connections to people across the world and chatting with them about shared interests. They were forming Facebook groups and using them to spread their ideas. That the chats were about hacked emails, and the pages and the groups were dedicated to promoting conspiracies about Clinton, was beside the point. Facebook made it easy for them to reach their intended audience.

The hackers also knew that the salacious details in the hacked emails were fodder for fringe websites and groups that would hype the Clinton-related material. The Russians doled out the emails strategically, for maximal impact. Just before the Democratic National Convention in July 2016, roughly twenty thousand emails from the DNC suddenly appeared on WikiLeaks. The emails showed DNC leaders playing favorites among the Democratic nominees for president. Most notably, DNC chairwoman Debbie Wasserman Schultz appeared to be pushing for Clinton over Bernie Sanders, the progressive senator from Vermont. The emails made front-page headlines, and Wasserman Schultz was forced to resign.

Another batch of emails, this time from Clinton campaign head John Podesta, were released just as the Trump campaign was suffering one of its worst embarrassments, an *Access Hollywood* tape that showed then–reality star Trump speaking disparagingly about kissing, groping, and grabbing women without their consent. The Podesta emails, which highlighted mudslinging within the campaign and included embarrassing revelations, such as of Clinton having been fed a question ahead of one of the town hall–style events, helped distract from the *Access Hollywood* tape, taking the heat, and the attention, off Trump and once again embarrassing the Clinton campaign. The Russian hackers had essentially become the world's most powerful news editors, enticing journalists to write stories with the promise of ever-more salacious emails from Democratic Party members. It was clickbait built for Facebook: difficult to ignore and primed for sharing. Just as news of one email began to die down, the hackers would have another leak ready to go.

The more Moran and the rest of the Facebook team probed, the more certain they were of the Russian link between the accounts. The hackers had been sloppy, at times forgetting to turn on the VPNs (virtual private networks) that would have masked their location, or leaving behind other traces that pointed to their locations in Russia and their connections to one another. Throughout July and August, Stamos and his team filed report after report about the Russian activity to Colin Stretch. In one, Moran described watching the DCLeaks Facebook page communicate with a journalist in real time about the hacked emails. The journalist, who worked for a right-wing publication, pressed for the emails to be sent as soon as possible, even in their raw form. And he took advice on how to "frame the story" around them.

Days later, a story appeared by the journalist that quoted extensively from hacked Clinton campaign emails. Moran felt responsible.

Before Moran's boss, Alex Stamos, arrived at Facebook, he had made a name for himself in the cybersecurity world for sounding the alarm at Yahoo. In April 2015, he threw open the doors of the company's offices in downtown San Francisco and invited several hundred journalists, cybersecurity experts, and academics to a conference he had named an "un-conference." The gathering was intended to point out failures to protect internet users, rather than to celebrate the newest technology often promoted at cybersecurity conferences.

At the time, Stamos was Yahoo's information security officer and one of the youngest and most high-profile cybersecurity experts in Silicon Valley. He had grown up in California's hacker community, a precocious coder with a degree in electrical engineering and computer science from the University of California, Berkeley. By age thirty-five, he had started and sold a successful cybersecurity company, iSEC Partners. Over the course of his career, he had been called in to consult with some of the most powerful companies in Silicon Valley as they uncovered Russian and Chinese hackers on their networks.

In most cases, the hackers had been there for years, and Stamos was appalled at how vulnerable companies remained to relatively simple attacks. "I wanted to point to the fact that the security industry had drifted away from what we all say our mission is, which is to keep people safe," he said. Companies didn't make security a priority, and they relied on overly complex systems that left them and their users vulnerable to attack. He used the analogy of a house: rather than create a solid, defensible structure and securely locking all the doors, he said, companies had done the equivalent of creating multistory residences with fragile windows. All a hacker needed to do was bide his time and wait, and eventually, a company would leave itself exposed.

Hundreds of people gathered for Stamos's keynote address in the nondescript conference room. Stamos weaved nervously through the crowd, making small talk in his characteristic rapid-fire fashion. At the lectern, he tucked in his red plaid shirt, adjusted the hem of his gray suit jacket, and jammed his hands into his pockets. People were still standing and chatting when Stamos, too eager to wait any longer, launched into his talk.

"I'm not happy," he began, "with where we are as an industry."

Technology companies weren't thinking of the privacy and security needs of the average person, he said. Cybersecurity companies were more interested in selling flashy and expensive technology than offering basic protections that small companies could afford. Each year, stories of hacks affecting millions of people were growing more common. People's private information, including Social Security numbers and credit card details, were increasingly being sold and traded by hackers in online forums.

Many in the room nodded along as he spoke. Stamos didn't disclose it in his speech, but he was growing concerned about the security practices of his own company. He had reason to be worried. Just weeks after the "unconference," his team found a vulnerability that made it possible to enter Yahoo's systems and search through a user's emails. Stamos rushed to check if Russian or Chinese hackers were responsible for the breach. Instead, he discovered that the vulnerability had been placed there intentionally, with the approval of CEO Marissa Mayer, to allow the U.S. government to monitor Yahoo email users secretly. "It was a huge breach of trust," he recalled. "I couldn't stand by and watch it." In less than a month, he would quit his job at Yahoo. When Mayer's role in granting the government secret access was made public, it would cause a national scandal.

In cybersecurity circles, Stamos became known as a "warrant canary." At the turn of the twentieth century, canaries had been sent into coal mines deep down below the earth's surface. If the birds died, it was a silent signal that toxic gas had been released into the air and that the shaft wasn't safe for miners. In the mid-2000s, the phrase was repurposed by internet companies. The companies had begun to receive secret government subpoenas for their data— the requests so secret that it was illegal even to reveal their existence. As a workaround to alert their users that their data was being accessed, internet companies began placing tiny images of yellow canaries or messages on their websites as a sign that all was well. If, one day, the canary disappeared, it meant the company had been served one of the secret subpoenas. Privacy experts and advocates knew to look for the canaries and to sound the alarm if one suddenly went missing.

When Stamos left Yahoo, it was a sign that something was deeply wrong there. His arrival at Facebook, however, indicated that he saw something at the company worth working for.

In the spring of 2015, Facebook executives were quietly looking around for a new head of security to replace Joe Sullivan, who was leaving for a similar post at the ride-hailing firm Uber. Zuckerberg and Sandberg asked Sullivan to help find his successor. Sandberg said she wanted a high-profile hire as a signal to their board of directors, and to regulators in Washington, of their commitment to security.

Sullivan suggested Stamos. He argued that Stamos had a history of holding the powerful to account: as an independent contractor, he had been notoriously fired or forced to resign by employers because of his fierce adherence to privacy and security. Sandberg expressed some concern, asking if she could trust Stamos to do what was best for Facebook. Sullivan demurred. He couldn't promise that Stamos would always fall into line, but there was no one else in the cybersecurity community with the same type of broad credibility.

When he was hired, Stamos was promised latitude to grow his team and to shape Facebook's approach to security. He would fall under Sandberg's remit, but she assured him that he would have the freedom to expand and structure the team as he saw fit. Under Stamos, Facebook's threat intel team expanded from a few members to over a dozen. He went on a more general hiring spree as well, doubling Facebook's entire security team from roughly 50 people to more than 120. Many of his hires had past government experience; he wanted to hire people trained in fighting the type of cyberwarfare he suspected was being waged on Facebook's platform.

The team Stamos built was unique at Facebook, both by design and location. His office, and most of his employees, sat in a far corner of the campus, in a building few employees had ever heard of, let alone visited. Building 23 was unadorned save for the big white numbers on its façade. Inside, it looked like most of Facebook's other offices, with minimalist bleached-wood furniture and glass-walled conference rooms lending it a clean, impersonal aesthetic. Stamos's security team filled their workspaces with gym bags, family photos, and throw pillows. Unlike the rest of Facebook's employees, who moved desks so frequently that they often refused to hang a single photograph, Stamos's team was given permanent desks, which they covered with thick binders of declassified intelligence reports and congressional briefings. They advertised their nerd chops by hanging hacker conference badges off their computer monitors. "Hack the Planet" posters decorated the walls. While the rest of Facebook drank IPAs and boutique beers at campus-wide happy hours, the security team had private cocktail hours with martinis and highballs mixed from an old-fashioned bar cart stocked with Johnnie Walker, Jose Cuervo, and a selection of mixers. Those who weren't based in Menlo Park, including Moran and other members of the threat intelligence team who worked out of DC, often joined by video.

Within a year of joining Facebook, Stamos had unearthed a major problem on the platform. But no one was responding to his reports, and the Russian activity was only escalating. On July 27, 2016, one of Facebook's security engineers watched from his sofa as cable news broadcast a live press conference of Donald Trump speculating about the DNC hack. Trump wasted no time zeroing in on his opponent, suggesting that Russia might have intercepted some of Clinton's emails from her private server. "I will tell you this: Russia, if you're listening, I hope you're able to find the thirty thousand emails that are missing," he said. "I think you will probably be rewarded mightily by our press."

The candidate's words left the engineer astonished. He opened up his laptop to check if news sites were reporting on what he had just heard. Had a U.S. presidential candidate just actively called for Russia to hack his opponent? The engineer walked to the shower, where he stood under the hot water for a long time. "It just felt," he said, "so wrong." All summer his company had witnessed the Russian hacking firsthand, and now Trump was cheering the hackers on to go even further.

At work later that week, he asked others on his team if they also felt sick to their stomachs. Most, including Stamos, shared his feeling of impending dread. "For those of us who did it professionally, we knew what Trump had just said wasn't a game," Stamos later recalled. "We already knew the Russians were engaged on some level. But we didn't know how much further they would go, especially when called to attack by a U.S. presidential candidate. We were really worried about how else Facebook could get used and manipulated."

Trump and the Russian hackers had separately come to the same conclusion: they could exploit Facebook's algorithms to work in their favor. Users could not resist the incendiary messages of populist politicians. Whether they agreed or disagreed, they engaged. Facebook's algorithms read that engagement as interest and silently tallied up points in favor of the content, which only boosted it higher and higher in users' News Feeds. The daily headlines about Trump's most recent preposterous claim or falsehood may have been intended as critical coverage by news organizations, but the never-ending drumbeat of front-page stories also meant that the candidate was impossible to escape, both on news sites and on Facebook.

Trump's team understood how to manipulate the platform through his Facebook page, with the candidate using it to post daily, and sometimes hourly, proclamations about the race. Meanwhile, his campaign was busily buying up millions of dollars in Facebook ads and navigating through Facebook's fine-tuned system to make sure its messages were constantly reaching the voters it needed most. It was also raising millions from its supporters through Facebook.

The Trump campaign's obvious dominance of the platform renewed a discussion that Facebook's employees had been quietly holding for years: did Facebook favor populists? Trump was only the most recent example. Over the last few years, the campaigns of Narendra Modi in India and Duterte in the Philippines had used Facebook to win over voters. Facebook employees in both countries raised their concerns with managers over the company's role in both elections. But it was not until Facebook found a populist leader rising in the United States that concern grew more widely across the company.

Stamos's team was growing frustrated. A dedicated team of Facebook lawyers passed the threat intel team's reports on to the FBI. But they were met with silence.

Stamos understood why Facebook was not going public with this, especially as the team was unsure what type of surveillance operation the U.S. intelligence agencies were conducting and did not have the type of evidence that would let them definitively attribute the Russian activity to that government in a court of law. But Stamos also felt a sense of responsibility. He told the team that he would take their concerns to Stretch and others in the legal department. Stamos raised it in a weekly meeting he held with the general counsel and was told that Facebook's legal and policy team took the concerns of the security team into consideration. Stamos couldn't shake the feeling that he was being brushed off, but he also knew that for Facebook to go out and name a country as an aggressor of a cyberattack would be unprecedented. "The norm of working through FBI was pretty well established," he said, adding that Facebook had a strong relationship with the law enforcement agency and had helped it prosecute many cases related to human trafficking and endangering minors. "There was no precedent before 2016, of a social media company coming out and doing direct attribution, where we publicly named a group that was responsible for hacking."

Within the threat intel group, there was debate over what should be done. Facebook was a private company, some argued, not an intelligence agency; the platform was not duty-bound to report its findings. For all Facebook knew, the National Security Agency was tracking the same Russian accounts the company was seeing, and possibly planning arrests. It might be irresponsible for Facebook to say anything. Others argued that Facebook's silence was facilitating Russian efforts to spread the stolen information. They said the company needed to make public that Russia-linked accounts were spreading hacked documents through Facebook. To them, the situation felt like a potential national emergency. "It was crazy. They didn't have a protocol in place, and so they didn't want us to take action. It made no sense," one security team member said. "It felt like this was maybe a time to break precedent."

In the fall, the threat intel team members saw the opening they needed. The Russian accounts they had been monitoring had turned their attention to a new type of campaign. The Fancy Bear hackers had stolen 2,500 documents by hacking into a foundation run by the billionaire hedge fund manager George Soros.

Soros's Open Society Foundations had been designed as a global philanthropic organization dedicated to pro-democracy values. Over the years, however, it had been falsely accused of many conspiracies; in November 2015, Russia's Office of the Prosecutor-General announced that Open Society was banned from Russia for posing a threat to state security. It was clear that Russia saw Soros and his organization as a threat.

Using the same infrastructure of Facebook pages and accounts they had put in place to share the Democratic Party emails, the Russian hackers attempted to convince journalists and bloggers to write about the hacked Soros documents. One of the security analysts on Facebook's security team, Jen Weedon, was paying close attention as the Russian hackers shared the documents they had accumulated. Within those documents was buried the personal information of Soros staffers. Weedon knew that sharing other users' personal information was strictly against Facebook's rules and was grounds for banning an account.

She approached Joel Kaplan with what she had found. Kaplan agreed that allowing DCLeaks to continue operating had left Facebook potentially legally liable; the only responsible action was to remove the page immediately. It was a victory based on a technicality, but many members of the security team breathed a small sigh of relief. Facebook still had no plan for reining in the abundance of Russian activity, but at least they had found a way to remove one of the primary culprits.

And yet, in the final months leading up to the vote on November 8, 2016, Facebook was inundated with vitriolic and polemical content that sought to divide Americans. Some of it was loosely based on the emails the Russians had hacked from the Clinton campaign; some was pure fiction. Blogs and websites with names no one had ever heard of before ran stories claiming that Clinton had fainted onstage, mothered a secret love child, or secretly ordered acts of domestic terrorism on U.S. soil. The stories were so outrageous that users couldn't help but click, even if only to try to find out who was behind them.

One engineer on Facebook's News Feed called it "a perfect storm." The week before the election, he was sitting at the Rose and Crown, a pub in downtown Palo Alto, with a half-dozen other friends who worked across Facebook's engineering teams. Over plates of fish and chips and draught IPAs, the group discussed what they were seeing at work. It was embarrassing, said one, to have to explain to family members that Facebook wasn't going to do anything about the conspiracies and false news being peddled across the platform. Another added that he had stopped telling people he worked at Facebook because all they did was complain.

"We laughed and told each other that at least it was almost over," said the engineer. "Clinton would win, and the Trump campaign would stop, and life would just go back to normal." It had never occurred to the group, he said, that Trump might win the presidency.

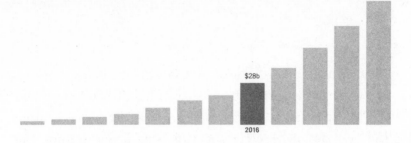

$28b

2016

Chapter 6

A Pretty Crazy Idea

At Facebook's lobbying office, exactly one mile from the White House, the policy team gathered at 9 a.m. on Wednesday, November 9, 2016. There was a long silence as everyone took their seats in a conference room. One employee had driven through the night back from Clinton's official "victory" party at the Javits Center in Manhattan. Others hadn't bothered to conceal their puffy, red-rimmed eyes.

Finally, Joel Kaplan spoke. He appeared as stunned as the rest of the team. Trump's victory put Facebook and the nation into uncharted territory, he said. He warned that Facebook would be blamed for the swell of misinformation on the site ahead of the election. Some DC employees privately thought the company deserved the blame, that its lax policing of fake news had in fact helped elect Trump.

We're in this together, Kaplan said.

He hadn't endorsed the candidate. Many of his friends were "Never Trumpers," Republicans who had taken vows to fight against Trump's candidacy. And of course, it didn't help that Sandberg, one of Clinton's "Hillblazers" (the campaign's designation for those who donated or raised at least $100,000 for the former secretary of state), had aligned herself so closely with the Clinton campaign.

The challenges in front of Facebook were clear. The company had built its influence operations in Washington to thirty-one lobbyists on the payroll, including internal and contract lobbyists. During Kaplan's tenure, the office had established a better balance of Democrats and Republicans, including Greg Mauer, former aide to Republican House Speaker John Boehner; and Kevin Martin, the former Republican chairman of the Federal Communications Commission. The Washington office, however, was still more focused on getting politicians to use the site than lobbying on particular issues. In the parlance of Washington, Facebook's goal was "defense" influence, to fend off regulations that threatened the growth of its business.

Kaplan made it clear he was ready to move to offense. Within three days of the election results, he had created an "administration pipeline," a list of people with ties to Trump whom Facebook should try to hire directly or as consultants. Among them were Trump's former campaign manager, Corey Lewandowski, and Barry Bennett, a senior adviser to the campaign, who toured the office late in 2016. The two men had formed the lobbying firm Avenue Strategies right after the election and were pitching themselves as Trump experts capable of decoding and explaining the president-elect's thinking on issues related to any matter of business and trade. Lewandowski promised that if Facebook put the new firm on retainer for $25,000 a month, he would use his access to Trump in its favor.

For many employees, it was surreal to see Lewandowski and Bennett walking the halls of the DC office. Kaplan had known Bennett, a friend and a familiar figure in Republican politics. But the combative Lewandowski, known for his sycophantic attachment to Trump, was new to national politics and lobbying and was a lightning rod for controversy. He had been accused of misdemeanor battery (a charge later dropped) for grabbing a reporter's arm during a rally. Disliked by many on the campaign, Lewandowski was fired weeks before Trump secured the official nomination from the Republican Party.

The office was ambivalent about hiring Lewandowski and Bennett. They were uncertain of Trump's position on tech. Some employees reasoned that the two men would be ideally positioned to help shield Facebook from the president's unpredictable targeting of individual companies. But others protested, saying Lewandowski did not fit the company's culture or brand. Ultimately, Kaplan was wary of Lewandowski, Washington staff said, and passed on the pitch.

In Menlo Park, Zuckerberg sat in the Aquarium, attempting to confront a different type of fallout from the election. In the days following the vote, a narrative had emerged that Facebook was to blame for Trump's surprise victory.

Zuckerberg ordered his executives to scramble. If the world was going to accuse Facebook of facilitating Trump's win through a deluge of false news and sensationalist stories, Zuckerberg wanted to counter with hard data. Facebook needed to prove to the world that false news was just a tiny sliver of the content users saw in their feeds. If he was able to show hard numbers, he said, the narrative would shift, and people would look elsewhere to understand Clinton's defeat.

While Facebook sat on a repository of data encompassing every article, photo, and video shared on the platform, and while it could, if it wanted to, calculate the exact number of minutes users spent looking at those pieces of content, no one at the company had been keeping track of false news. Cox, the head of product, along with the head of News Feed, Adam Mosseri, now ordered engineers to determine the precise percentage of fake news on the platform. With many of the conference rooms booked up days, even weeks, in advance, the execs held meetings in the cafeteria and hallways.

It was trickier than anyone anticipated. There were shades of gray between stories that were outright false and those that were misleading. The team bandied around some numbers and concluded that whatever figure they came up with would fall within the single digits. But when they presented their findings, executives quickly realized that if even only 1–2 percent of content on Facebook qualified as false news, they were talking about millions of stories targeting American voters ahead of the election. Word was sent up to Zuckerberg that the numbers he was looking for weren't ready; it would take weeks before they had solid answers for him.

In the meantime, Zuckerberg had committed to appearing at a conference on November 10, less than forty-eight hours after Americans had voted. When he took to the stage at the Techonomy conference at the Ritz-Carlton at Half Moon Bay, twenty miles from MPK, it was inevitable that he would be asked about the election. His press team advised him to take the question at the conference, where the interview was guaranteed to be friendly and relaxed. The stage was set with just two chartreuse chairs—one for Zuckerberg, the other for David Kirkpatrick, the founder of the conference. In 2010, Kirkpatrick had published the first book chronicling the rise of Facebook; he had met extensively with Zuckerberg both before and after the book's publication. Facebook's press team saw him as a friend of the company and hoped he would cast Zuckerberg's answers in as positive a light as possible. It was, in the words of one Facebook PR person, "as soft an interview as we could hope for."

As expected, Kirkpatrick started the interview with a question about the election, wondering if Facebook's bird's-eye view of the world had given it any sense that Trump might win the presidency. "You could measure the number of Trump-related posts vis-à-vis Hillary, and the relative enthusiasm could be measured. I can imagine ways that you could figure that out?"

"We could do that," Zuckerberg answered. "I can rattle some stuff off the top of my head here. It's not news that Trump has more followers on Facebook than Hillary does. I think some of his posts got more engagement. Anyone could have looked at this before." But Facebook, Zuckerberg insisted, hadn't spent time analyzing those figures. And the company could not easily re-create what appeared in someone's News Feed, due to a thousand different factors that determined what users saw every day. Facebook, said Zuckerberg, had seen the news reports accusing the company of littering users' News Feeds with junk. He wanted to dismiss outright the notion that those stories had had any effect on how the electorate voted.

"I think the idea that fake news on Facebook—of which it is a very small amount of the content—the idea that influenced the election in any way is a pretty crazy idea," Zuckerberg said. His voice grew stronger as he continued. "I do think there is a certain profound lack of empathy in asserting that the only reason why someone could have voted the way they did is because they saw some fake news."

The statement, taken out of context, made headlines around the world. *Facebook feeds were inundated with false news.* And now Zuckerberg, just days after the election, was utterly dismissive of the potential consequences. It wasn't credible. Commentators on CNN and NBC lambasted Zuckerberg and his company for failing to take responsibility for their role in the election.

Facebook employees were confused. Several of them posted the link to Zuckerberg's talk on their Tribe boards and asked for feedback from their coworkers. Was Zuckerberg being truthful? they asked. Hundreds of employees responded, in messages spun off the original posting and in smaller chats among groups of friends.

Some vowed to press their managers for answers and to insist that the company conduct an audit or hire an outside firm to investigate the role that false news had played in the election. Others just kept repeating the question: had Facebook incentivized a network of false and hyper-partisan content in such a way that the company had helped make Donald Trump the next president of the United States?

Stamos was in Germany, holding meetings about the coming European elections and the threat of meddling on social media. He saw Zuckerberg's "crazy" quote in a news story and then searched for the complete interview to make sure the reporting was accurate. He read the full transcript and was stunned. Zuckerberg seemed to have no idea that his security team had uncovered a range of alarming Russian activity on the platform. For months, Stamos's team had been filing reports to his immediate superiors, and he had assumed that Zuckerberg and Sandberg were being briefed. Stamos had had no contact with the company's top executives himself.

That week, he asked for a meeting with Zuckerberg and Sandberg.

$41b

2017

Chapter 7

Company over Country

"Oh fuck, how did we miss this?" Zuckerberg asked, looking around at the somber faces in the Aquarium. In silence, the group of nearly a dozen executives stared at Alex Stamos, hoping he had the answer. Stamos, clutching his knees as he perched on the end of Zuckerberg's couch, was perspiring in his merino wool sweater.

It was the morning of December 9, 2016, and the group had gathered to hear a briefing by Stamos outlining everything the Facebook security team knew about Russian meddling on the platform. Due to security concerns, Stamos had not emailed his findings to the attendees ahead of time. Instead, he printed out handouts for each person in the room.

"We assess," the summary read, "with moderate to high confidence that Russian state-sponsored actors are using Facebook in an attempt to influence the broader political discourse via the deliberate spread of questionable news articles, the spread of information from data breaches intended to discredit, and actively engaging with journalists to spread said stolen information."

The eight-page document laid out how Facebook's security team had, in March 2016, first spotted Russian activity on the platform by hackers collecting intelligence about key figures in the U.S. elections. It explained that Russians had created fake Facebook accounts and pages to spread disinformation and false news. It provided a history of Russia's military intelligence agency, or GRU, and an overview of its capabilities. It suggested that Russia's next target for election interference was the May 2017 election in France, in which the right-wing candidate, Marine Le Pen, was challenging Emmanuel Macron. The Facebook team had already spotted Russian operatives preparing a major effort to sway French voters. It bore a striking resemblance to the operation they had just conducted in the United States, including false Facebook accounts primed to spread hacked materials to embarrass Macron and bolster Le Pen.

The report concluded with a section entitled "Looking Forward," which acknowledged that Facebook was sitting on a trove of information proving that a foreign government had tried to meddle in the U.S. election. "Most of what is discussed in this document is not yet public, as few organizations have the visibility we have into government activity on social media," it read. "However, with a bipartisan push in Congress for open hearings and President Obama ordering a review of Russian activity during the election, it is possible that Facebook will be publicly identified as a platform for both active and passive information operations."

The threat was only increasing, the report made clear: "We expect that our challenge with organized disinformation campaigns will only increase in 2017."

No one else spoke as Zuckerberg and Sandberg drilled their chief security officer. Why had they been kept in the dark? How aggressive were the Russians? And why, asked a visibly agitated Sandberg, had she not known that Stamos had put together a special team to look at Russian election interference? Did they need to share what Stamos had found with lawmakers immediately, and did the company have a legal obligation beyond that?

Stamos was in a delicate position. Schrage and Stretch, who had been receiving his reports for months, were seated directly across the table, but neither man spoke up, and it didn't seem wise to blame them, he thought. There were also legal considerations at play, Stamos realized. His investigation could expose the company to legal liability or congressional oversight, and Sandberg, as the liaison between Facebook and Washington, would eventually be called to DC to explain Facebook's findings to Congress. "No one said the words, but there was this feeling that you can't disclose what you don't know," according to one executive who attended the meeting.

Stamos's team had uncovered information that no one, including the U.S. government, had previously known. But at Facebook, being proactive was not always appreciated. "By investigating what Russia was doing, Alex had forced us to make decisions about what we were going to publicly say. People weren't happy about that," recalled the executive. "He had taken it upon himself to uncover a problem. That's never a good look," observed another meeting participant.

For his part, Schrage felt strongly that Stamos had still not discovered enough to determine Russia's involvement definitively. "I left that meeting feeling that Alex [Stamos] had raised more questions than answers—how significant were the fake news clusters on Facebook? How could we remove them? Were they motivated by financial or political objectives? If the motivations were political, who was behind them?" Schrage recalled. "More digging was necessary before Facebook could know what policy changes or public announcements to make."

Stamos felt that he had been trying to sound the alarm on Russia for months. In order even to get in front of Zuckerberg and Sandberg, he had first been required to present his assessment to their top lieutenants. At that meeting, held the day before, he had been scolded by Boz and Cox for not coming to them directly when his team first flagged the Russian interference. Both men enjoyed relative autonomy within their departments, but that power was contingent on their running Facebook smoothly—Zuckerberg trusted them to be his eyes and ears and to carry out his mission. It hadn't occurred to Stamos that he needed to loop them into the investigation, however; he reported to Stretch. "It was well within my remit to investigate foreign activity within the platform. And we had appropriately briefed the people in our reporting chain," he said. "It became clear after that that it wasn't enough. The fact that the product teams were not part of that conversation was a big problem."

Boz had accused Stamos of purposefully keeping the Russian investigation secret. Cox had wondered why his team hadn't been looped in sooner. Stamos replied that he had no direct line of command to either of them. He could see that he had put too much faith in Facebook's reporting structure. He was still new to the company and had barely interacted with Zuckerberg or Sandberg. "It was my own failing, that I hadn't made it into the inner circle of trust with them. Maybe someone else, with a different personality, could have done better than I did in getting their ear," Stamos realized. "I was on the outside." Boz and Cox didn't like what Stamos had exposed them to, and they were embarrassed by what they didn't know. But they knew this was big. Schedules were cleared to meet with Zuckerberg and Sandberg the next day.

Now, in the Aquarium, Stamos gave a somber assessment of where they stood, admitting that no one at the company knew the full extent of the Russian election interference. Zuckerberg demanded that the executives get him answers, so they promised to devote their top engineering talent and resources to investigate what Russia had done on the platform. But with the Christmas and New Year holidays looming, it would take nearly a month to assemble a team of security experts across the platform.

In the weeks before he was sworn into office, Trump received briefings by the intelligence community and outgoing members of the Obama White House that were unequivocal about Russia's election meddling. In public, however, Trump had waffled about Russia's involvement. While he conceded in one interview that Russia had been behind the hack, he denied in others that the Russians had had anything to do with his victory. He spoke about Russia, and Putin, in warm, even glowing terms. Facebook's Washington lobbyists had delivered a clear message to Menlo Park: the suggestion that Russia had done anything to influence the election was not well received by the Trump administration and would be seen as a direct attack on the president himself. There were also concerns about how the company would be viewed if it took action to remove false news from the platform. Conservatives increasingly believed Facebook was working to suppress their voices. While there was little evidence for their claims—repeated studies showed that conservative voices had, in fact, become increasingly popular on Facebook over the years—the allegations were stoking antagonism within the Trump administration.

At the close of the meeting, Zuckerberg and Sandberg agreed that Stamos and others should redouble efforts to get to the root of the Russian election interference. The new group formed to investigate if Russia had used Facebook to meddle in the election—which included Stamos and a handful of his security experts, engineers, and advertising employees and was led by one of Facebook's earliest employees, Naomi Gleit—began to meet in the Menlo Park offices. They called themselves "Project P"—the P for "propaganda." The group met daily and discussed its progress on a secure Facebook page. The banner image was a scene from a 1950s movie adaptation of the George Orwell classic *1984*, the dystopian novel that depicts a future in which much of the world lives under constant government surveillance. The irony was thick. Project P was trying to find propaganda disseminated through Facebook's platform, and they were the ones doing the surveillance. "Big Brother Is Watching You" became the group's slogan, written at the top of its Facebook page, an inside joke among Facebook's security staff.

By late January, when Project P was ready to wrap up its work, it had not found much more on the Russian campaign. But it had uncovered a huge network of false news sites on Facebook. Some were run by Macedonian teenagers, others by American dads working out of their basements. The users shared a range of articles: a mix of hyper-partisan reports and pure works of fiction. Some of the articles claimed that Hillary Clinton had secretly passed away in the midst of her campaign and had been replaced by a doppelganger; others suggested the "deep state" was secretly propping up the Clinton campaign. The motives were financial. Each click equaled money that the sites made through ads.

On February 7, 2017, Stamos drafted a briefing for Facebook's policy team that laid out the project's findings. He closed the report with a recommendation: "We need a dedicated effort to anticipate, understand and implement imaginative, cross-functional solutions." It was essential, in Stamos's view, that the company prepare for another potential attack. Yet, for Facebook's management team, the main takeaway appeared to be relief that no major Russian intrusion had been found. "There was a feeling that maybe we could close the book on this. At least, that is how Sandberg described it," a Project P team member recalled. "There was a feeling that maybe Stamos's warnings had been overblown."

Stamos and other members of the security team—working under what was known as an XFN, or "cross-functional," group—continued to search for Russian interference and broadened their scope to look for disinformation campaigns affecting other countries. The security team had already found that governments were actively using the platform to promote their own political agendas. In a separate report produced by the XFN team, Stamos included flags from nations such as Turkey and Indonesia as examples of governments that had used Facebook to run disinformation campaigns to sway public opinion and elections in their countries or those in nearby states.

Facebook needed to go on the offensive. It should no longer merely monitor and analyze cyber operations; the company had to gear up for battle. But to do so required a radical change in culture and structure. Russia's incursions were missed because departments across Facebook hadn't communicated and because no one had taken the time to think like Vladimir Putin.

Stamos delivered his team's reports to his bosses and shared it with Schrage and Kaplan. He knew that they were key to winning Sandberg's support: if they approved of a project or idea, it virtually guaranteed the COO's endorsement as well. But Schrage and Kaplan were not ready to make big changes at the company, according to one member of the policy team who was privy to the discussions. "They saw Stamos as alarmist. He hadn't uncovered any big Russian interference, and what he was suggesting with disinformation seemed hypothetical. They also knew that if they took it to Sheryl and the company announced a big re-org, it could draw more scrutiny and attention from outsiders."

Schrage and Kaplan still did not think Stamos had collected enough data to be able to gauge the scope of the Russian interference. Nobody could say whether what the Russians had done on Facebook had successfully influenced American voters ahead of the election; why alarm the American public? In discussions with colleagues, they raised the point that by drawing attention to the security team's findings, they could also invite unwanted attention by lawmakers in Congress. As a private global company, Facebook did not want to engage in geopolitical skirmishes, and it certainly didn't want to be in the middle of contentious national elections. Facebook was, above all, a business.

It was a line of thinking that came directly from Zuckerberg. In Facebook's earliest days, when their office was still a glorified loft space, "Company over country" was a mantra the CEO repeated to his employees. His earliest speechwriter, Kate Losse, wrote that Zuckerberg felt that his company had more potential to change history than any country—with 1.7 billion users, it was now in reality already larger than any single nation. In that worldview, it made sense to protect the company at all costs. Whatever was best for Facebook, whatever continued the company's astronomic growth, user engagement, and market dominance, was the clear course forward.

Disclosing that Russia had meddled in the U.S. elections served no purpose for Facebook, but it performed the civic duty of informing the public and potentially protecting other democracies from similar meddling. Facebook's security team had already seen what type of impact they could have in France. In December, they had watched as a Facebook account that went by the name "Leo Jefferson" began probing the accounts of people close to the French presidential elections. Tracing that account, the team discovered that Russian hackers were attempting to replicate during the French presidential elections what they had done in the United States.

They quickly reached out to intelligence officials in Paris, and within days they had delivered a warning for France. Still, Russian hackers released a cache of twenty thousand emails from the Emmanuel Macron campaign on Friday, May 5, just two days before France was set to vote. Facebook's warning of Russian involvement helped intelligence officials quickly declare the hacked emails a Russian plant, and news stories reported that people should not trust what they read in the emails. Macron went on to win the election. It was a quiet moment of victory for Facebook's security and threat intel teams, staffed largely by people who had public service backgrounds. In the United States, though, they were finding themselves increasingly at odds with the company's objectives.

In April 2017, Stamos and his team had pushed to publish a white paper, an in-depth report for the public on security concerns facing the platform, including the Russian interference they had discovered during the elections. In its first iteration, the paper contained an entire section devoted to activity by state-backed Russian hackers. It detailed how a network of Russian actors had probed Facebook accounts through "spear phishing"—sending a malicious file or link through a seemingly innocuous message—and collected intelligence on them. It provided concrete examples from the 2016 presidential elections that demonstrated how Russia had used the platform to spread hacked documents. Stretch agreed to show the white paper to Schrage and other Facebook executives. The word came back: Remove the Russia section. Facebook could not risk going public with its conclusion that Russia had interfered in the 2016 elections, Stamos was told. The management team considered it politically unwise to be the first tech company to confirm what U.S. intelligence agencies had discovered. "They didn't want to stick their heads out," said one person involved in the discussions.

Stamos worked with two threat intel staffers on variations of the Russia section. They deleted the details, describing things more broadly. They took out mention of the election, and Stretch, Schrage, and Kaplan, who was asked to give his assessment from DC, reviewed the new draft. Again, word came back that they should eliminate any mention of Russia. Stamos and his team were reminded that Facebook was not a government or intelligence agency; it was not its place to insert itself in international relations.

Sandberg did not participate in any of the conference calls. "Sandberg didn't get directly involved, but everyone on the security team knew she had been looped into the decision and that Schrage was just the messenger on what she wanted," observed one security team member. Zuckerberg was aware that the white paper was being drafted, but he did not weigh in or review the drafts. His 2017 New Year's resolution was to visit all fifty U.S. states. In carefully crafted posts featuring photo-ready appearances astride tractors and on factory room floors, Zuckerberg shared his meetings with the American Everyman, a chance to show he cared about the country. But they also frequently took him away from the office. It was clear to Stamos's team that neither Zuckerberg nor Sandberg was focusing on the security team's work.

On April 27, 2017, while Zuckerberg was in Dearborn, Michigan, visiting the Ford automotive factory and learning how to assemble a car, Facebook released the white paper. The word "Russia" never appeared; the only vestige was a footnote linking to a report by U.S. intelligence agencies that cited Russia's meddling in the 2016 elections.

Many on the security team were surprised and angered by the omission. At least one member of the team challenged Stamos, telling his colleagues that Stamos had not advocated forcefully enough for their findings. He accused his boss of being bulldozed by Facebook's policy and legal teams. "There was a sense that Stamos could have gone and knocked down walls to get the word out on what Russia had done," recalled the security team member. "But that would have been dramatic, and probably not accomplished anything. We started to feel like we were part of a cover-up at Facebook."

Stamos seemed to agree. He began to speak to those closest to him on the team about stepping down. "I'm not sure I can do the job anymore," one colleague recalled him saying.

Sen. Mark Warner sat in Sandberg's office in late May 2017, surrounded by posters exalting the mottos of *Lean In*. Earlier that year, Warner had become vice chair of the Senate Intelligence Committee, one of the most powerful committees in Congress. The position afforded him broad oversight into a number of intelligence issues, and since assuming that role, he had become increasingly convinced that Facebook was involved in Russia's attempts to meddle in the election. He had met several times with Stamos and members of Facebook's policy team in Washington, but had repeatedly walked away with the feeling that the company was holding something back.

Now he was in Menlo Park to confront Sandberg directly. The sofas, where Sandberg preferred to sit for meetings, were ignored, and the group sat formally at the conference table. Warner had brought his aides, and Sandberg had assembled her staff, as well as Stamos. It was clear that Warner had an agenda, and his questions for Sandberg and Stamos were pointed—he wanted to know what evidence Facebook had uncovered on Russia's election interference.

Earlier that month, *Time* magazine had published a story that alleged that U.S. intelligence officials had evidence that Russia had purchased Facebook ads as a way of targeting U.S. voters ahead of the 2016 presidential elections. Stamos and other members of the security team had read the piece with frustration. For months they had been searching for Russian ads, but due to the sheer volume of business, it was like looking for a needle in a haystack. If the U.S. intelligence community had a lead, why had they not shared it with Facebook? During his meetings in DC, Stamos had come away with the feeling that members of Congress weren't coming clean. No one had said a word about evidence surfaced by U.S. intelligence agencies.

They had reached an unspoken impasse: Warner felt that Stamos was not giving him full answers on what Facebook knew about Russia, and Stamos believed Warner was withholding information from intelligence agencies about Russia's capabilities. Both men were wrong.

As they sat across the conference table, Sandberg weighed in. "We are wrapping this up," she told Warner, locking eyes with the senator. She gave him a reassuring smile as she promised that Facebook would come to his office with anything new. The company, she said, was confident it had found the vast majority of the Russian activity.

Stamos was taken aback. While it was true that he had shared the bulk of Project P's findings with Warner, his team suspected there was much more to be uncovered and were still digging into the Russian activity. Even as they met, his threat intelligence team was searching for the Russian ads. He felt he had to hedge Sandberg's positivity.

He gently added to his boss's assessment. Facebook, he said, had uncovered a lot, but it was not sure that other Russian groups weren't still active. Stamos turned directly to Warner and asked once more for assistance: "Can you help us, share with us anything you know?" Warner demurred but promised to keep in touch.

Stamos left the meeting feeling doubly concerned. Clearly, Sandberg wasn't being briefed on the real-time operations of his team. Worse yet, she had just told a sitting senator, one of the most powerful in Washington, that Facebook considered its investigation into Russian election interference complete. He dashed off an email to the threat intelligence team stressing the importance of developing a system to track down the Russian ads. If U.S. intelligence officials were unwilling to provide Facebook with their information, Facebook would have to search for the needle in the haystack itself.

In Washington, the threat intelligence team resumed their search in the only way they knew: through trial and error. Jen Weedon, Ned Moran, and others were painstakingly sifting through Facebook's immense ad trove. They began by reviewing every advertisement run in the year leading up to the November 2016 elections and compiling a list of every ad that contained political messaging. They also ran searches for ads that mentioned the candidates by name and their running mates. Those searches yielded hundreds of thousands of results.

They needed first to determine which, if any, of the ads had been bought by someone in Russia. The group searched for all ads bought by accounts that had Russian set as their default language. Another search targeted ads that were bought by Russian-based accounts. Yet another returned all ads that had been paid for using rubles.

The group also used the platform's access to detailed account metadata. They saw, for instance, if the person who bought the ad had accidentally forgotten to turn off their location information on a previous occasion and had popped up with an IP address located in Russia. Facebook also had the ability to check the phone or computer from which an advertisement was purchased. In many cases, the ads appeared to have been purchased through phones that had Russian SIM cards installed, even if they were using technology to try to mask their location. In other cases, the phones appeared to be secondhand, and even though the current SIM cards in the phone had no location data, Facebook saw that the phones had previously been used with Russian SIMs.

It was Moran who found the first group of advertisements that appeared to have been purchased from Saint Petersburg. He had been bringing his work home with him, searching late into the night. The rest of his team was accustomed to his singular focus and the way in which he could seemingly go whole days without speaking as he analyzed the data in front of him. It was due precisely to this intense focus that he began noticing the accounts tied to Saint Petersburg: the name had stuck in his memory because of a *New York Times* exposé written in 2015 about a Russian group known as the Internet Research Agency, or IRA. In the story, the reporter, Adrian Chen, had traveled to Saint Petersburg and documented Russia's efforts to use a troll army stationed at the IRA to shape pro-Russian narratives across the internet. Moran scanned the article again to check, and sure enough, Saint Petersburg was listed as the headquarters for the IRA.

He slowly began to build on his hunch, and by the middle of June, he felt he had collected enough examples of accounts to prove a pattern. He had managed to identify the first cluster of Facebook accounts that were based in Saint Petersburg, which were promoting political ads aimed at Americans and working in concert with one another. It took him a full day to check and double-check his work, as he reran all his searches to make sure he had not made any mistakes. By the time the searches were complete, he was home and had just finished putting the baby to sleep.

His first call was to Stamos. "I think we found something, and I think it is Russian," he said.

Stamos asked him to run through the evidence tying the accounts to Saint Petersburg at least twice, then suggested a few loose ends that Moran could tie up to see if there were more clues linking the accounts. But it was clear that Moran had found the needle.

When he got off the phone, Stamos called his wife and told her they would need to cancel their upcoming vacation.

With Moran's find, Facebook's legal and security teams quickly put together a search protocol based on a new classifier. Facebook engineers had built these before, when the company was looking for sexual predators, violent extremists, or other groups they wanted to weed out from the platform. The engineering team had become adept at building the classifiers, but they had to be given precise search parameters.

They also needed to greatly expand their search. Up to this point, the security team had been looking at a small sample of ads that were overtly political; they now needed to comb through every ad that had ever run on the platform. Much of Facebook's data was locked away in storage in data centers spread out across the United States. At times, it took weeks to locate the facility in which the hard drives were stored. In many cases, data was misplaced, and the factory workers had to spend hours searching through cavernous warehouses for the precise box with the servers Facebook's security team needed.

Once they had the data, they needed to start tying the clusters of ads back to the IRA in a way that was definitive. Working with Isabella Leone, an ads analyst, Moran had connected his initial cluster to Saint Petersburg, but they needed to establish more ties between the ads they were finding and the IRA. A newly hired analyst, Olga Belogolova, and a longtime security analyst, April Eubank, began to pore over the data. Both women had expertise in Russia's tactics and were skilled at deep data dives. Their eureka moment came the first week of August, when Eubank, who had been running a data query, excitedly called the rest of the team over to her desk. She had a graph open on her monitor that showed numerous lines connecting accounts that had bought advertisements to accounts used by the IRA. Each line represented a connection, and there were many of them.

Over the course of six weeks, the security team constructed a map of the Internet Research Agency's activity on the platform. The Russians had bought more than 3,300 advertisements, which had cost the IRA roughly $100,000 to promote on Facebook; some of those advertisements did little other than promote the 120 Facebook pages the IRA was running. They also produced more than 80,000 individual pieces of organic content. In total, an astonishing 126 million Americans had been reached.

Wherever there was a "seam line issue," a divisive position that could turn Americans against one another, there was the IRA. Gun control, immigration, feminism, race—the IRA ran accounts that took extreme stances on each of these issues. Many of their pages supported the Trump campaign and conservative groups across the United States, but they also ran pages in support of Bernie Sanders. "It was a slow, slow process, but what we found that summer, the summer of 2017, it just blew us away," recalled one member of Facebook's security team. "We expected we'd find something, but we had no idea it was so big."

While over the years some Facebook employees had speculated that the IRA was focused on spreading disinformation in the United States, no one had thought to go looking for a professional disinformation campaign run by the organization. The security team had not believed the IRA audacious or powerful enough to target the United States. Facebook's assurances to Senator Warner and other lawmakers that they had fully uncovered Russia's efforts to influence the 2016 elections were based on the assumption that they were looking for Russian military intelligence activity, but they had missed the forest for the trees. While they were chasing state-backed hackers spreading emails from the Clinton campaign, they had overlooked the IRA's highly effective troll army.

The discovery of the IRA ads prompted Stamos's team to double down. In Menlo Park, they were already isolated in their building on the outskirts of the Facebook campus; Moran and a small number of other members of the threat intelligence team were working in a dedicated room in Facebook's DC offices. They were all instructed not to discuss their findings with any of their colleagues. They were told that executives wanted to produce a comprehensive report on the IRA activity before they went public with their findings.

So, the team was forced to watch in silence as Facebook's own spokespeople unknowingly misled the press. On July 20, by which time some clusters of Russian ads had already been found but had not been linked directly to the IRA, a DC-based Facebook spokesperson informed a CNN anchor that the company had "seen no evidence that Russian actors bought ads on Facebook in connection with the election." The spokesperson worked less than one hundred feet from where the threat intelligence team sat, but he had not been briefed on its findings.

On August 15, "Project Campaign," a dedicated team that had been formed following the discovery of the ads, published an internal report that highlighted hundreds of Facebook and Instagram accounts that had been linked to the IRA. The report revealed that the team had also found, but had not analyzed, real-world Facebook events created by the IRA, such as protests and other gatherings, as well as a number of IRA-run pages that were co-run by Americans. The IRA had even sent emails and messages to Facebook employees, asking to learn more about the company's policies and have their pages and accounts reenabled when they ran into technical problems.

Project Campaign's report recommended that, going forward, Facebook immediately change its policies to allow the security team to remove any Facebook account found participating in "coordinated inauthentic behavior"—defined as any accounts that "mislead users about the origin of content on Facebook," "mislead users about the destination of links offsite," and "mislead users in an attempt to encourage shares, likes, or clicks."

The summer was coming to an end, and Congress had returned from its August recess primed for sharper inquiries into the Russian election meddling. Zuckerberg and Sandberg had been briefed on the security team's findings in weekly meetings. The Facebook executives knew they had to go public, but they wanted to do it on their terms.

They decided that the company would come clean with everything it knew the day before the quarterly board meeting scheduled for September 7. Board members were not aware of the work the security team had conducted over the summer, or of the IRA campaign the team had unearthed. To avoid leaks and to control the press, members of Congress and the public would be briefed on the sixth. It would not be easy, they thought, but at least Facebook could try to control the narrative and convince the public that it had finally uncovered the extent to which Russia had interfered in the 2016 elections.

Stamos and his team were asked to prepare a security briefing for the board and to write a blog post for Facebook. Having Stamos's name on the blog served a dual purpose. His team had been responsible for finding the Russian activity and knew it best, and his reputation as a warrant canary might win over the cybersecurity world.

Schrage and Facebook's legal team eviscerated the first draft of the post, which Stamos had crafted in coordination with several members of his team. It should be as academic as possible, Stamos was told, and carefully worded to ensure that Facebook did not open itself up to more legal liability than was necessary. "He was told to take the most conservative numbers possible, and present them," recalled one person involved in the conversations. "Everything he had written, the context around the investigation and the accounts the security team had found, all of that was eliminated from the blog." It felt like a repeat of what had happened with the white paper—once again, Facebook was withholding what it knew.

Stamos hoped that before the board, at least, he could present a full picture of what his team had uncovered. He repeatedly asked Schrage if he should prepare a deck, or some type of reading material, for the board members ahead of time, but he was told to keep his presentation limited to documents he could present in person.

On the morning of the sixth, Stamos was escorted to a soundproof room to appear before a special, three-member audit committee whose members were pulled from Facebook's full board and were dedicated to governance and oversight, including security and privacy issues. The three members were Marc Andreessen; former chief of staff to President Bill Clinton, Erskine Bowles; and Susan Desmond-Hellmann, an oncologist and chief executive officer of the Bill and Melinda Gates Foundation. They sat at a table with Stamos at the head; Schrage and Stretch were also present. Over the course of an hour, using a deck he printed out and distributed to the three board members, Stamos ran through the security team's findings.

The board members remained largely silent, speaking only when they wanted Stamos to slow down or explain something in greater detail. When Stamos was done, Bowles reacted first. "How the fuck are we only hearing about this now?" The profanity was softened by his lilting Southern accent, but there was no mistaking his anger. The unflappable North Carolina patrician rarely cursed in the Facebook boardroom. The other two board members reacted with similar, if tempered, fury. Andreessen grilled Stamos on the technical aspects of Russia's playbook. Desmond-Hellmann wanted to know more about how the security team had gone about finding the accounts and what it planned to do next.

Stamos gave his honest assessment. It was likely that there was more Russian activity on Facebook that his team had not yet discovered, and it was almost guaranteed that Russian attempts to influence U.S. voters were ongoing.

The audit committee recommended that the full Facebook board be shown the presentation Stamos had prepared and that the subject of Russian election interference be raised to the top of the full board meeting, planned for the next day. It did not go well. "There was definitely yelling. There were questions of 'How could you let this happen?' and 'Why are you only telling us now?'" reported one person who was briefed on the meeting. Zuckerberg maintained his calm, walking the board through some of the technical solutions he hoped to put in place, which included a hiring spree among the safety and security teams. Sandberg was more shaken up. "At the end of the day, Stamos's team reported to Sandberg, and the policy teams reported to Sandberg. There was this feeling that she had dropped the ball."

Sandberg was especially upset by Stamos's assessment that there could be millions of dollars of ads, or hundreds of Russian accounts, still active on the platform. Those concerns had not been mentioned in the reports and briefing notes she had seen, and in her mind, Facebook had found the majority of what there was to find on the platform. Yes, what had happened was regrettable, but she hoped the board would walk away with the sense that Facebook had uncovered the full extent of the Russian interference. When told by members of the board that Stamos had presented a very different picture in the meeting the day before, raising the possibility that there were Russian ads and accounts that had not yet been found, she froze.

The next day, in a hastily called meeting in the Aquarium, she took out her frustrations on the chief security officer. "You threw us under the bus!" she yelled at Stamos, in front of dozens of people gathered in the room as well as others who had joined the meeting via a conference line. The outburst continued for several minutes, with Sandberg upbraiding Stamos for taking so long to find the Russian interference and for concluding, without any evidence, that there was still more. Stamos shrank in his seat. He did not offer a response or defense, and no one else in the room said a word until Zuckerberg, visibly uncomfortable, said, "Let's move on."

Facebook had bigger problems to contend with than the board meeting. In the hours after Stamos had met with the audit board, the company had published the blog post detailing Facebook's findings. Innocuously titled, "An Update on Information Operations on Facebook," it was the first public acknowledgment of IRA activity on the platform, reporting that not only had Russia tried to sway the U.S. elections, but it had spent approximately $100,000 on ads from June 2015 to May 2017. The roughly three thousand ads it had run had violated Facebook's policies, wrote Stamos, and the accounts associated with them had been removed. The post omitted the fact that millions of Americans had been reached through those ads.

Stamos knew the post did not give the full picture of Russia's election interference. He was unsurprised when the figures he'd provided were immediately torn apart by academics and independent researchers who had long been studying Russia's presence online.

In San Francisco, disinformation researcher Renée DiResta reviewed the Facebook blog post from her apartment overlooking the bay. Almost immediately, she began texting researchers and amateur disinformation sleuths she had developed relationships with over the last two years.

The number of people actively studying disinformation in the United States could be listed in one breath. DiResta, like the others, had chanced into studying the field. When researching preschools for her firstborn son in 2014, she discovered that a number of preschools in Northern California had lax rules on vaccinations.

DiResta wanted to know why so many parents were casting off the recommendations of the medical establishment and so began joining anti-vaccine groups on Facebook.

In her twenties, DiResta had studied market dynamics, drawing on her love of data-heavy charts and graphs to understand patterns within the financial markets. She applied that same approach to study the way anti-vax groups shared information online. The activists were using Facebook to recruit people to their movement in unprecedented numbers. Facebook's own algorithms were their greatest asset. Once someone joined one group touting "natural health cures" or "holistic medicine," they were led down a rabbit hole where, within a click or two, they were invited to join an anti-vax group. Once a user was there, Facebook recommended scores of other anti-vax groups for them to join.

The groups were lively, posting multiple times a day about dramatic stories of children allegedly negatively affected by vaccines. As always, all the engagement was interpreted by Facebook's algorithms as a sign of popularity, and the stories found their way to the top of users' News Feeds. DiResta began charting how the groups shared information, noting how they coordinated their announcements and events with one another to achieve maximal impact on the platform. She began to publish her research—and quickly found herself under attack by anti-vaccine activists.

DiResta's research established her as someone who understood how information was shared on social networks. In 2015, she found herself consulting with an Obama-led White House team studying how ISIS used Facebook and other social media networks to recruit for its terrorist network and spread its propaganda online. ISIS, concluded the White House, used the same tactics as the anti-vaccine activists. Facebook, DiResta thought, had built the perfect tool for extremist groups, be they anti-vaccine activists or ISIS, to grow their numbers and their influence online.

In the wake of the September 6 blog post about the Russian ads, DiResta was quickly sought out by members of Congress, who wanted her help in assessing what Facebook had found. She told them that in order to understand the extent of what the company had uncovered, she and other researchers needed access to the data itself. She wanted to see the Facebook ads and analyze the Facebook pages used by the Russians. Soon, lawmakers were demanding the same. During phone calls with Facebook lobbyists in the days after the announcement, legislators and their staffers demanded that the company hand over its data. When Facebook's lawyers declined to share the content, the government officials went to the press with the complaint that Facebook was withholding evidence.

Facebook executives realized that they would need to answer the calls from Congress, or risk getting into a protracted news cycle over the ads. "Facebook's Washington office told them that no matter what, it was a lose-lose situation. If they didn't give up the ads, Congress would be furious and never let them forget it. If they did, they risked setting a dangerous precedent and potentially angering Trump," recalled one member of Facebook's Washington office.

But with DiResta and other researchers quickly finding many of the ads on their own, Sandberg and Zuckerberg decided that it was time for them to find a way to work with Congress.

Not long after the blog post about Russian interference came out, two Facebook lobbyists entered the basement library of the House Permanent Select Committee on Intelligence in the Capitol Building, a windowless, wood-paneled room lined with shelves of books on law enforcement and national intelligence. They were greeted by Democratic and Republican investigative aides for the committee. They came to present, for the first time, examples of Russian-bought ads.

Adam Schiff, the ranking Democrat of the committee, had been bearing down on the company to disclose what it had discovered and reveal how Russian actors had used Facebook content to manipulate voters during the presidential race. "One hundred thousand dollars may seem like it's not a huge amount, but at the same time, that's millions of people seeing or liking or passing on disinformation, and that can be influential," Schiff pointed out after reading the blog post earlier that day.

Greg Mauer, Facebook's designated lobbyist for House Republicans, and Catlin O'Neill, his equivalent for House Democrats, presented the aides with a manila envelope and told them they could look at the ads but couldn't keep them. Inside were about twelve images printed on standard 8-by-11 paper that the company representatives described as indicative of the kinds of ads they had found. The sampling suggested that the ads targeted Republican and Democratic candidates in equal measure. There were negative and positive ads on Bernie Sanders, Hillary Clinton, and Trump.

The committee staff were incredulous. In their own investigation, they had seen that the Russians appeared to be pushing Trump as their candidate of choice. Mauer and O'Neill pointed to an ad featuring Clinton with a woman in a hijab; "Muslims for Clinton" appeared below the photo in a font meant to mimic Arabic script. The company representatives, the Democratic congressional aides said, had presented it as a promotion the Russians had run in *favor* of the candidate. The aides, however, interpreted it differently: the ad was clearly meant to stir up ethnic tensions and to give the impression that Hillary Clinton was supported by radicals, they thought. If this was the one example Facebook had of a pro-Clinton ad, it was not convincing. "We were like, 'Are you kidding me?'" recalled one of the aides in the meeting. "I suspect they were aware of the political tornado they were entering into and probably wanted to present a case that wouldn't anger either side. The decision to present this very small subset of ads that they described as both sides of the narrative, when it clearly was not, didn't inspire trust."

That decision had, in fact, been reached after fierce internal debate on how to provide the most minimal amount of information that would appease angry members of Congress. Since the blog post went public, Facebook's press and policy staff had been going in circles formulating a response. Tom Reynolds, a congressional staffer turned Facebook spokesperson, was getting hourly queries from reporters at the *Washington Post*, the *New York Times*, CNN, and other outlets. He had predicted that the press would hit them hard. In an email sent on the afternoon of September 6, he warned his colleagues that the media were questioning Facebook's line that they could not share content from the pages or ads. "As we move into the next chapter of coverage, reporters will push more on 'Why are you not releasing ad content, or impressions/reach of ads?'" The email, which went to members of the security, policy, and press teams, explained that Facebook spokespeople were telling reporters that they did not share the ads due to federal laws protecting users' data. While the legal issues were real, Reynolds argued, Facebook was going to "get pushed more on these topics."

Molly Cutler, a Facebook lawyer, laid out the company's legal reasoning once more, reiterating that Facebook should hew to the party line that it was acting out of concern for users' privacy. Brian Rice, a Democratic lobbyist with Facebook, replied that the response would not cut it with Congress or the public. "I don't think we should rely on our terms and privacy obligations," he wrote. "Regardless of what is required of us by law, the easy jump will be 'Facebook protects privacy of Russian trolls at the expense of Western Democracy.'"

There were endless ways to interpret the privacy question. Some on the legal team argued that sharing the Russian-made content was a violation of the Electronic Communications Privacy Act, which was designed to prevent the government from getting access to private electronic communications. Handing over the content and ads the Russians had created was, according to these lawyers, a violation of its users' privacy and set a precedent for Congress asking Facebook to hand over content from other private individuals in the future. But Facebook's policy and security teams felt that by following the letter of the law, Facebook was missing the spirit of the law.

On September 21, Zuckerberg spoke about the Russian ads publicly for the first time, in a Facebook Live video. He would work with Congress in its investigation and turn over the ads, he said. "I don't want anyone to use our tools to undermine democracy. That's not what we stand for."

Within days, the company sent a lobbyist to deliver a thumb drive to an aide on the House committee with the first batch of the 3,393 ads. On October 2, Schiff announced that the committee had received the ads and would make a sampling of them public. Both Senate and House Intelligence Committees sent out press releases announcing that they would hold hearings with executives of Facebook, Twitter, and Google on Russian interference in the elections. "The American people deserve to see the ways that the Russian intelligence services manipulated and took advantage of online platforms to stoke and amplify social and political tensions, which remains a tactic we see the Russian government rely on today," Schiff said.

The files in the thumb drive provided the bare minimum of information: two- to three-page PDFs with some basic metadata on the first page such as account ID; how many views, shares, and likes the ad had received; and how much the ad had cost. The additional pages featured images of the actual ad. The PDFs did not include geolocation or more detailed metadata; Facebook's policy staff and lawyers had also negotiated with the committee to redact certain images and pieces of content that the IRA had stolen or copied from real people. Those people were innocent bystanders, the House committee staff and lawyers agreed, and exposing them would be a violation of their privacy.

Three committee aides spent three weeks sifting through the files. They found that many were unreadable because of the amount of content blacked out by Facebook. The congressional aides argued with Facebook to lift some of the redactions, file by file, at least for the committee to review. When Facebook agreed, the aides found many of the redacted images and text were of commercial images or were well-known internet memes. "It was incredibly lazy and frustrating," one aide said.

While Facebook battled with lawmakers in private, Sandberg revved up the public relations operation. The COO came to DC on October 11, three weeks ahead of the hearings, to meet with lawmakers and reporters. She began the day with a live-streamed interview by Mike Allen, the high-metabolism political journalist and founder of the new digital news site Axios, a must-read for politicos. "Things happened on our platform that shouldn't have happened" in the lead-up to the 2016 presidential election, Sandberg conceded. "We know we have a responsibility to prevent everything we can from this happening on our platforms."

During a private meeting a few hours later with Schiff and a representative from Texas, Mike Conaway, the leading Republican on the committee, Sandberg promised that Facebook was taking the issue seriously. "She was there to catch arrows and to protect her left flank," as one Facebook official put it. But she also made clear that the company wouldn't disclose what it had found. She told Schiff that it was Congress's job to inform the public.

Sandberg's DC tour did little to dampen concerns among lawmakers. On November 1, after months of oblique references to the divisive Russian-backed ads, the House Intelligence Committee showed samples of the ads during its hearing with the general counsels for Facebook, Twitter, and Google. Legislative aides enlarged images of the ads and mounted them on poster boards propped up around the hearing room.

Schiff's opening remarks got straight to the point. "Today, you will see a representative sample of those ads, and we will ask the social media companies what they know about the full extent of Russian use of social media, why it took them so long to discover this abuse of their platforms, and what they intend to do about it to protect our country from this malign influence in the future." Schiff and Conaway then directed attention to the ads displayed around the room. One ad, from an account called "Being Patriotic," showed a photo of six police officers carrying the casket of a fallen colleague. It was captioned, "Another Gruesome Attack on Police by a BLM Movement Activist." Another, from "South United," depicted a Confederate flag with the accompanying text "Heritage, not hate. The South will rise again!" The ad, created in October 2016, cost approximately $1,300 and garnered roughly forty thousand clicks. Some representatives in the hearing noted that Trump campaign and administration officials, including Kellyanne Conway and Michael Flynn, had shared some of the ads.

Schiff also provided an overview of the reach of Russia's IRA on the platform. Based on information provided by Facebook, he said, the committee had concluded that a total of 126 million Americans were likely exposed to content from an IRA page, and 11.4 million had viewed an ad placed by the troll farm.

When called to the witness table, Facebook's general counsel, Colin Stretch, promised that the company was learning from its mistakes. It planned to hire more people on the security team and was using AI to find inauthentic accounts. But he also acknowledged that the IRA had been able to achieve wide reach: "They were able to drive a relatively significant following for a relatively small amount of money. Its widest activity appears so pernicious, it was undertaken, I think, by people who understand social media. These people were not amateurs, and I think [this] underscores the threat we're facing and why we're so focused on addressing it going forward."

Facebook's security team felt like they were under attack. They had found a Russian disinformation campaign that no one, including the U.S. government, had predicted or thought to look for. But instead of being thanked, they were being mocked for missing the campaign and for the delay in making their findings public. Stamos, especially, felt under siege. He took to Twitter to complain about the press's interpretation of the Russian ads, but he found his comments met with skepticism. His reputation as a warrant canary was tarnished. The cybersecurity community wanted to know why he had kept quiet for so long about the evidence his team had uncovered.

He focused his attention on the company. He wanted to make sure Facebook did not repeat the same mistakes again. Over the course of the fall of 2017, he prepared several proposals for restructuring Facebook's security team. One of Facebook's talking points in the immediate aftermath of the September 6 blog post had been that it was expanding security manpower; in a September 11 email, a spokesman, Tom Reynolds, had argued that telling reporters that Facebook was expanding and upgrading the security team would "give us something positive to point to." Yet, in the months that followed, Stamos had no clear sense of whether Facebook was actually considering hiring more security team members. He was also uncertain of the status of another idea he had proposed in October: hiring an independent investigator like Ash Carter, the former secretary of defense under Obama, to examine how Facebook had handled the Russia investigation.

By December, Stamos, losing patience, drafted a memo suggesting that Facebook reorganize its security team so that instead of sitting off on their own, members were embedded across the various parts of the company. If everyone was talking to one another, he argued, it left less chance that another elaborate information offensive like what the Russians had already plotted would fall through the cracks. For several weeks, he heard nothing back from his bosses. It wasn't unusual; with the Christmas holidays approaching, many people were distracted. He decided he would reopen the topic when everyone had returned to work in January. Then, in the midst of his Christmas break, he received a call from Stretch.

Facebook had decided to take his advice, but rather than organizing the new security team under Stamos, Facebook's longtime vice president of engineering, Pedro Canahuati, was assuming control of all security functions. Left unsaid was what Stamos's new role at the company would be. The decision felt spiteful to Stamos: he had advised Zuckerberg to cut engineers off from access to user data. No team had been more affected by the decision than Canahuati's, and as a result, the vice president of engineering told colleagues that he harbored a grudge against Stamos. Now he would be taking control of an expanded department at Stamos's expense.

Stamos was shocked and angry. "I got frozen out. I was really upset. Mark accepted the plan that eviscerated my team without talking to me about it," he later recalled.

When he returned to work in January, the security team of over 120 people that he had built was largely disbanded. As he had suggested, they were moved across to various parts of the company, but he had no role, or visibility, into their work. Instead, he was left in charge of a whittled-down team of roughly five people.

The sole moment of celebration came on the afternoon of February 20, when Special Counsel Robert Mueller announced that he had indicted or secured guilty pleas from thirteen people and three companies during an investigation that encompassed election interference charges against overseas Russian nationals and crimes committed by Trump advisers in the United States. In their offices in Washington, someone snapped a photo of Moran, Weedon, Eubank, Belogolova, and William Nuland, another threat intel team member, reading over the indictment at their computers and toasting one another's work.

"It was the first, public acknowledgment of what we had done, the piece of history that we had contributed to," said one member of the team. "It felt amazing to recognize that, even if it was just to ourselves."

Stamos had already handed in his resignation letter to Stretch, but had been persuaded to stay through the summer. Facebook, he was told, did not want a high-profile departure. He was told to think of his remaining security team members and was offered a lucrative compensation package if he stayed quiet and announced his resignation as an amicable parting of ways in August.

During the remainder of the time Stamos spent at Facebook, he didn't speak with Sandberg or Zuckerberg about the Russian disinformation campaign or his departure. There was no company-wide good-bye party. Instead, he gathered with the closest members of his erstwhile security team and drank tequila at the bar they had built in the corner of their secluded office building. The bar had been given a name by the team, and even after Stamos's departure, it was known as "Bar Serve and Protect."

$56b

2018

Chapter 8

Delete Facebook

On March 17, 2018, the *New York Times* and the *Observer* of London broke front-page stories about a company called Cambridge Analytica that had obtained profile information, records of likes and shares, photo and location tags, and the lists of friends of tens of millions of Facebook users. A whistleblower within the UK-based political consulting firm had brought the story to the news organizations with a stunning claim that the firm, funded by Trump supporter Robert Mercer and led by Trump's senior adviser, Stephen K. Bannon, had created a new level of political ad targeting using Facebook data on personality traits and political values.

But the jaw-dropping detail was that Cambridge Analytica had harvested the Facebook data without users' permission. "The breach allowed the company to exploit the private social media activity of a huge swath of the American electorate, developing techniques that underpinned its work on President Trump's campaign in 2016," the *New York Times* reported. The *Observer* wrote that the "unprecedented data harvesting, and the use to which it was put, raises urgent new questions about Facebook's role in targeting voters in the US presidential election."

It was the latest breach of trust in Facebook's repeated pattern of data privacy abuses. The company's long history of sharing user data with thousands of apps across the internet had opened the door for Cambridge Analytica to harvest data on up to 87 million Facebook users without their knowledge. But the case struck a particular nerve because of the firm's most famous client: the campaign of Donald J. Trump. With outrage over Facebook's role in election interference running high and the United States fiercely divided over Trump's election, the story brought together two raging veins of anger within the nation in a watershed privacy scandal.

Three weeks later, Zuckerberg sat at a small witness table in the cavernous wood-paneled hearing room of the Hart Senate Office Building. He wore a slim-cut navy suit and a tie of Facebook blue. He appeared fatigued, his face drained of color and his eyes sunken, he stared ahead unflinching as photographers jockeyed for position around him, cameras whirring in his face. An entourage of Facebook executives—Joel Kaplan, Colin Stretch, and several lobbyists—sat behind the CEO, wearing grim expressions.

"Facebook is an idealistic and optimistic company," Zuckerberg said in his opening remarks. He presented a benevolent picture of the social network as a platform that had spread awareness of the "MeToo" movement and helped student organizers coordinate the March for Our Lives. After Hurricane Harvey, users had raised more than twenty million dollars on Facebook for relief, he added. "For most of our existence, we focused on all the good that connecting people can do."

It was Zuckerberg's first appearance before Congress, and he faced a hostile audience. Hundreds of spectators, lobbyists, and privacy advocates had gathered outside the hearing room in a line that snaked around the marble halls. Protestors crowded the entrance to the building with "Delete Facebook" signs. Cardboard cutouts of Zuckerberg wearing a "Fix Facebook" T-shirt dotted the Capitol Building lawn.

Inside, he faced forty-four senators seated in tall black leather chairs on a two-tier dais. On his table, a small microphone was attached to a box with a digital countdown clock lit in red. His black leather notebook was open to his talking points—"Defend Facebook, Disturbing Content and Election Industry (Russia)"—next to a single yellow pencil.

Sen. Dick Durbin, a senior Democrat from Illinois, peered down at Zuckerberg over black-framed glasses perched on the end of his nose. "Mr. Zuckerberg, would you be comfortable sharing with us the name of the hotel you stayed in last night?" Durbin began.

Zuckerberg struggled for a response, glancing at the ceiling and laughing nervously. "Um no," he answered with an uncomfortable smile.

"If you messaged anybody this week, would you share with us the names of the people you've messaged?" Durbin continued.

Zuckerberg's smile began to fade. It was clear where the questions were headed.

"Senator, no, I would probably not choose to do that publicly here," Zuckerberg replied in a serious tone.

"I think this may be what this is all about," Durbin said. "Your right to privacy, the limits of your right to privacy, and how much you give away in modern America in the name of, quote, connecting people around the world."

How Cambridge Analytica breached Facebook users' privacy traced back eight years, to when Zuckerberg stood on the stage at the F8 developers' conference in San Francisco and announced the creation of "Open Graph," a program that allowed outside app developers to gain access to Facebook users' information. In return, Facebook got users to extend their sessions on the site. Zuckerberg then invited gaming, retail, and media apps to plug into Facebook, and in the first week after the F8 conference, fifty thousand websites had installed Open Graph plug-ins. Facebook offered the apps access to its most valuable asset: users' names, email addresses, cities of residence, birth dates, relationship details, political affiliations, and employment history. In the race to sign up partners, salespeople had been instructed to dangle the offer of data as an incentive. Safety and privacy were an afterthought. For every ten Facebook employees tasked with recruiting new partners to the Open Graph system, and helping them set up their systems to receive data from Facebook, there was one person overseeing the partnerships and making sure the data was used responsibly.

At least one Facebook employee, a platform operations manager named Sandy Parakilas, had tried to warn the company about the dangers of the program. In 2012, Parakilas alerted senior executives, including Chris Cox, that the program was a privacy and security stink bomb, and presented them with a PowerPoint that showed how Open Graph left users exposed to data brokers and foreign state actors. It was highly likely that Open Graph had spurred a black market for Facebook user data, Parakilas had cautioned. But when he suggested that Facebook investigate, the executives scoffed. "Do you really want to see what you'll find?" one senior official asked. Disillusioned, Parakilas left the company months later.

In 2014, Zuckerberg shifted strategies and announced that he was shutting down the Open Graph program because it wasn't yielding the engagement he wanted. Just before Facebook closed its doors to outside developers, though, an academic at Cambridge University named Aleksandr Kogan plugged into the Open Graph and created a personality quiz called "thisisyourdigitallife." Nearly 300,000 Facebook users participated. Kogan harvested data from those individuals as well as their Facebook friends, multiplying his data set to nearly 90 million Facebook users. He then turned the data over to a third party, Cambridge Analytica, in violation of Facebook's rules for developers.

The backlash to the scandal was swift and fierce. Hours after the story landed, Democratic senator Amy Klobuchar of Minnesota called for Zuckerberg to testify before Congress. The hashtag #DeleteFacebook started trending on Twitter. Republican senator John Kennedy of Louisiana joined Klobuchar on March 19 with a letter to the Judiciary Committee chairman asking for a hearing with Zuckerberg. Celebrities amplified the #DeleteFacebook hashtag, with Cher announcing the next day that she was deleting her Facebook account, which had been an instrumental tool for her charity work. There were "things more important" than money, she tweeted. Even friends of Sandberg and Zuckerberg emailed them saying they had deleted their accounts. Thousands of users tweeted screenshots of the messages they had received from Facebook confirming deactivation.

On Tuesday, March 20, 2018, Facebook's deputy general counsel, Paul Grewal, held an emergency meeting with all employees to inform them that the company had begun an internal investigation of Cambridge Analytica. Zuckerberg and Sandberg didn't attend, a red flag that generated more concern among employees.

That same day, members of the Federal Trade Commission informed reporters that the agency was conducting an investigation of its own into violations of Facebook's 2011 privacy consent decree. The consent decree had been a big deal—a sweeping settlement agreement over charges that the company had repeatedly deceived users and abused their data. Reached after months of negotiations, that settlement had resolved a complaint involving eight counts of illegal activity, starting with the change to privacy settings in December 2009. To settle the charges, Facebook had agreed to two decades of privacy audits, which the Democratic-led

FTC described as "historic," a new bar for privacy enforcement. The company had also agreed to inform users of any changes to its privacy policy. "Facebook's innovation does not have to come at the expense of consumer privacy," then–FTC chairman Jon Leibowitz had proclaimed in the official announcement. "The FTC action will ensure it will not."

And yet, here they were, seven years later, in apparent violation of the decree.

British authorities also opened an investigation into Facebook and seized Cambridge Analytica's servers. The United Kingdom had already begun to investigate the political consulting firm over its alleged role in the 2016 referendum in which Britons voted to leave the European Union, a move known as "Brexit." The new revelations about Cambridge Analytica's harvesting of Facebook data fanned concerns in Great Britain over political targeting in the lead-up to that contentious vote.

Facebook's stock had dropped 10 percent since the news broke, wiping out $50 billion of the company's market capitalization. As the crisis unfolded, Zuckerberg and Sandberg didn't make any public statements or appearances. News outlets questioned their absence. Vox, CNN, and the *Guardian* ran stories with the same headline: "Where's Mark Zuckerberg?" The answer wasn't terribly surprising. Zuckerberg and a group of his direct reports and PR experts had been camped out in the Aquarium, returning to their homes only for a change of clothes and a few hours of sleep. They took up every seat on the gray L-shaped couch and every one of the conference table chairs. The air in the room was stale, and paper coffee cups, soda cans, and candy wrappers littered the blond-wood table and floor.

Sandberg's conference room down the hall was in similar disarray. She had corralled her bleary-eyed army of policy and communications staff as well as her "kitchen cabinet" of external advisers, including David Dreyer, a former senior official for the Treasury Department, and his business partner, Eric London. The two ran a Washington, DC–based consulting and public affairs firm that Sandberg used as personal and professional consultants. The advisers were dialed into the room by speakerphone.

The first order of business, as Zuckerberg saw it, was to play catch-up. He ordered staff to shut down external communications until he had a grasp of the situation. He next directed Sandberg and the legal and security teams to scour emails, memos, and messages among Facebook employees, Kogan, and Cambridge Analytica to figure out how the company had lost track of its own data. But the employees who knew about the arrangement with Cambridge Analytica had either left Facebook or lost contact with their business partners. On March 19, Facebook hired a digital forensics firm in London to try to access Cambridge Analytica's servers. The UK Information Commissioner's Office turned the staff away; it had already seized the servers.

The little that Facebook knew was damning. The company had learned about Cambridge Analytica in December 2015, from a report in the *Guardian* on how the presidential campaign of Ted Cruz had hired the political consulting firm for its ability to use Facebook data to target voters. One of Facebook's partnership managers had ordered the company to delete the data after the story was published, but no one had followed up for confirmation.

Zuckerberg fixated on technical details, such as how Cambridge Analytica had obtained the data and how it was transferred from Facebook to Kogan and then from Kogan to Cambridge Analytica. The breach stemmed from his side of the business, the product side, and he was the architect of the Open Graph. Still, privately, he fumed at Sandberg. During their regular Friday meeting, he snapped that she should have done more to head off the story, or at least to control the narrative around it. As far as he was concerned, it was the second time she had failed to control the press narrative around a Facebook scandal; he had also taken note of her failure to spin the Russian election interference and the fallout in Congress. Sandberg quietly told friends at the company about their exchange and voiced her concern that Zuckerberg might fire her. They assured her that he was just letting off steam. The expectation that Sandberg could, or should, be held responsible for the crisis was unrealistic, some employees felt. "Cambridge Analytica came from a decision in the product organization that Mark owned. Sheryl has been an adviser and would say something is good or bad, but the decision rested with Mark," one former employee observed.

Even before the Cambridge Analytica scandal broke, Sandberg was finding the company's slow-motion fall from grace personally devastating. In January 2018, she had made her regular pilgrimage to Davos, where Trump spoke, but she stayed out of the spotlight. She didn't speak on any panels, but Facebook was constantly invoked by others. In a panel called "In Technology We Trust?," Salesforce CEO Marc Benioff noted that election interference and privacy violations demonstrated the need to regulate tech companies. That same afternoon on CNBC, he told journalist Andrew Ross Sorkin that Facebook was as dangerous as tobacco. "Here's a product, cigarettes, that are addictive, you know they're not good for you," Benioff said. "There's a lot of parallels."

Two days later, at the annual dinner he hosted each year, George Soros delivered a speech in which he expressed deep concern about the role of propaganda and disinformation campaigns on social media in the U.S. elections. A Hungarian-born Jewish refugee and survivor of the Holocaust, Soros laid partial blame on Facebook's business model, which he claimed hijacked attention for commercial purposes, "inducing people to give up their autonomy." He called Facebook and Google monopolies with neither "the will nor the inclination to protect society against the consequences of their actions."

The criticisms felt like a pile-on; no one was heralding the good the company did. Organizers for the 2017 Women's March in Washington had used Facebook to coordinate the event; the company had helped sign up two million Americans to vote; users tapped Facebook friends to find organ donors—the powerful examples of Facebook's contributions to society were getting washed out by a few bad examples of harm, Sandberg complained to aides. She liked to recount the story, now well known, of how Wael Ghonim, an Egyptian activist, had used Facebook to organize protests against the authoritarian regime under Hosni Mubarak, helping to spark an Egyptian Revolution that swept the country and spread to Tunisia. In an interview with CNN's Wolf Blitzer after Mubarak resigned, Ghonim exclaimed, "I want to meet Mark Zuckerberg one day and thank him." But Sandberg never spoke about how free expression also included harmful content—such as the conspiracy theory that the mass murder at Sandy Hook Elementary School was a fabrication or that vaccines would lead to autism. It was as if she curated her worldview to exclude negative or even critical feedback. When the company had relocated to the new Menlo Park offices, she had even named her conference room "Only Good News." The uplifting stories were validation both of Facebook's mission and of her ambition to run a company that was more than a profit center.

Sandberg responded defensively to attacks on the intentions of the company's leaders. To her, the grousing was reminiscent of the public backlash to *Lean In*. Leading into the interview portion of the *60 Minutes* segment featuring the book, Norah O'Donnell described Sandberg as a public school kid from Miami who had rocketed through Harvard, the Clinton administration, and the top ranks of Google to become one of the world's most powerful executives as the second-in-command at Facebook. But that was not what was putting her in the headlines now, O'Donnell added.

"In a new book that has already touched a nerve, Sandberg proposes a reason for why there are so few women at the top," the anchor said, raising a perfectly arched eyebrow. "The problem, she said, might just be women themselves."

Lean In had grown out of the commencement address Sandberg delivered at Barnard College in 2011, a call for women to lean into their careers that became a viral video and turned her into a new feminist icon. But O'Donnell put her finger on the growing resentment of not just the book but also its author. Sandberg, some thought, was unfairly blaming women instead of pointing a finger at cultural and institutional gender biases. She was letting men off the hook and asking women to succeed by men's rules. And perhaps most damning of all, she was out of touch with the reality of the majority of working women, women without partners at home to share domestic duties or the financial resources to hire nannies, tutors, and cleaners, as Sandberg could.

"You're suggesting women aren't ambitious," O'Donnell said.

The majority of the twelve-minute piece had presented a glowing profile of Sandberg, but when O'Donnell confronted her with the critiques of the book in the interview portion of the segment, Sandberg's expression tightened.

She wanted women to know that they had many options and shouldn't cut off professional opportunities, she quickly countered. The data showed that women weren't asking for raises or promotions or leading meeting discussions. It was her responsibility to share her research and life experience. Her intention, in writing her book, was to empower women and help them get smarter on how to navigate their careers. "My message is not one of blaming women," she said. "There's an awful lot we don't control. I am saying that there's an awful lot we can control and we can do for ourselves, to sit at more tables, raise more hands."

She had been warned that she would draw criticism for the book. One friend told her she came across as elitist: most women would not be able to relate to her experience as a white, highly educated woman with great wealth. But Sandberg was blindsided by the naysayers. Maureen Dowd, the *New York Times* columnist, described her as a "PowerPoint Pied Piper in Prada ankle boots reigniting the women's revolution." The COO might mean well, Dowd allowed, but her call to arms was based on an out-of-touch conceit. "People come to a social movement from the bottom up, not the top down. Sandberg has co-opted the vocabulary and romance of a social movement not to sell a cause, but herself."

To Sandberg, critics like Dowd missed the point. "My entire life, I have been told or I have felt that I should hold back on being too successful, too smart, too lots of things," she said to O'Donnell. "This is deeply personal for me. I want every little girl [to whom] someone says you're bossy to be told instead, you have leadership skills."

The backlash felt personal and unfair—an affront. She felt attacked on what she saw as unassailable territory: her good intentions.

Likewise, the public lashing of Facebook felt unwarranted. The company had become a scapegoat, a convenient target for business rivals, Sandberg told senior staff in the wake of the Cambridge Analytica scandal. The media portrayals of the company were born from jealousy: newspapers blamed the decline of the publishing industry on Facebook, and the media were punishing the platform with negative coverage. Other executives thought there was an even simpler explanation for the backlash: that if Cambridge Analytica weren't associated with Trump, there wouldn't be a controversy. "Trump's election is why everyone is mad at us," one longtime executive insisted.

Someone needed to make a public stand. Five days after the scandal broke, Zuckerberg agreed to let a CNN reporter into the Aquarium for an interview. The CEO began on a familiar note of contrition. "This was a major breach of trust, and I'm really sorry that this happened," he said, wearing his deer-caught-in-the-headlights look. The company would begin an audit of all the apps that could have possessed and retained sensitive data, he assured CNN's Laurie Segall.

But when Segall asked why Facebook didn't make sure back in 2015 that Cambridge Analytica had deleted the data, Zuckerberg bristled. "I don't know about you," he said, not bothering to hide his impatience, "but I'm used to when people legally certify they're going to do something, that they do it."

Zuckerberg had prepped for his congressional appearance with a team of litigators from the DC law firm WilmerHale. In mock hearings, they bombarded him with questions about privacy and election interference and quizzed him on the names and backgrounds of each of the lawmakers. They warned him about questions out of left field intended to shake him off his talking points.

The stakes were high: the hearings promised to pull back the curtain on Facebook's cash cow of targeted advertising and force Zuckerberg to defend a part of the business he rarely discussed in public. The CEO was famously nervous in public appearances, prone to breaking into a sweat and stuttering through hard interviews. His staff rarely challenged him, so executives weren't sure how he'd respond to tough questions, grandstanding, and interruptions—hallmarks of congressional hearings.

Facebook's lobbyists had fought to prevent Zuckerberg from testifying. Sandberg was the company's appointed representative to Washington, and she was an unflappable and reliable public speaker, never deviating from the company message. At Trump's meeting with tech executives soon after his election, Sandberg had represented Facebook at the Trump Tower gathering, along with the CEOs of Amazon, Apple, Google, and Microsoft. (Kaplan, who accompanied Sandberg, stayed one day longer to interview with the Trump transition team for the position of director of the Office of Management and Budget. He withdrew before a decision was made.) But lawmakers had refused to accept a substitute for Zuckerberg. When Dick Durbin and Marco Rubio, Republican senator from Florida, threatened to subpoena Zuckerberg, the Facebook founder's staff agreed to the marathon two days of hearings before more than one hundred lawmakers. As a small concession, they asked congressional aides to crank up the air-conditioning for the CEO.

Some of Facebook's lobbyists and communications staff watched a livestream of the hearing in a conference room in the Washington offices. In Menlo Park, executives gathered in a glass conference room with the hearing displayed on one TV screen and their colleagues in DC on videoconference on another. The employees winced as lawmakers took shots at Zuckerberg for repeated privacy violations over the years and his empty apologies to the public. Many of the questions took direct aim at Facebook's business model. Public relations staff sent texts to reporters in the hearing room to gauge their impressions of Zuckerberg's performance. The consensus seemed to be that he was steady and cool, unfazed by interruptions or by the recriminations of wrongdoing. He was clearly prepared and able to answer a wide range of questions.

For the most part, Zuckerberg stuck to the script. But he also dodged multiple questions by promising that his staff would follow up with answers. And he delivered Facebook's standard defense that it gave users control over how their data was used and that the company didn't barter data for profit. "We do not sell data to advertisers. What we allow is for advertisers to tell us who they want to reach. And then we do the placement."

Also watching the hearing was Jeff Chester, the privacy advocate. He stood in his small home office in Takoma Park, Maryland, in a fury. Scattered across the floor around him were piles of Facebook's brochures and PowerPoints for advertisers that showcased a different view of the company—materials Chester had collected at advertising conferences where Facebook bragged about the power of its ad tools to global ad agencies and brands like Coca-Cola and Procter and Gamble. Inside the hotel ballrooms of ad industry conferences like Advertising Week in New York, Facebook's executives boasted about the company's unparalleled stockpile of data and its ability to track users off the site. They claimed it had more data than any other company and that it was able to help advertisers influence the minds of its 2.2 billion users. But to the public, Facebook rarely talked about the business in that manner.

Chester watched now as Sen. Roger Wicker, the chair of the Commerce Committee, asked if Facebook tracked users when they browsed other sites. Facebook had a tool called "Pixel" that allowed it to collect data on its users while they were off the site, a tool well known in the ad and tech industries. But Zuckerberg dodged the question. "Senator, I want to make sure I get this accurate, so it would probably be better to have my team follow up afterward," he said.

Chester couldn't contain himself. Facebook itself promoted such products to its advertisers. He tweeted a link to Facebook's marketing material on Pixel and other tools it used to track users while they were not on the platform. For the next three hours, every time Zuckerberg gave vague or misleading answers, Chester tweeted links to evidence of Facebook's data mining and behavioral tracking. The Facebook CEO was deliberately misleading the clueless committee members, Chester thought. He had been complaining to journalists for years about how Congress, both parties, had allowed Facebook to mushroom into a "digital Frankenstein," and now that they had Zuckerberg at the witness table, they were bungling it.

Facebook was at its heart an advertising company. By 2018, along with Google, it constituted a duopoly in digital advertising, with a combined $135 billion in advertising revenue reported in 2017. That year, Facebook's advertising revenue surpassed that of all U.S. newspapers combined. The platform's powerful tracking tools could follow users off the site and had gathered user data that could be broken down into more than fifty thousand unique categories, according to an investigation by the nonprofit news site ProPublica. An advertiser could target users by religious preference, political leaning, credit score, and income; it knew, for instance, that 4.7 million Facebook users were likely to live in households with a net worth of $750,000 to $1 million. "They started an arms war for data. For better or worse, Facebook is an incredibly important platform for civil life, but the company is not optimized for civil life," explained Ethan Zuckerman, the creator of the pop-up ad and an associate professor of public policy, communication, and information at the University of Massachusetts at Amherst. "It is optimized for hoovering data and making profits."

For years, Chester and other consumer advocates had warned regulators that Facebook was pushing new boundaries and thriving in the rules-free environment at the expense of consumers: it was harder to get government approval for a radio license in rural Montana or to introduce a new infant toy, they pointed out, than to create a social network for a quarter of the world's population. Over the past two decades, Congress had proposed multiple laws to protect online privacy, but under tremendous lobbying pressure by tech companies over details in the bills and gridlock in Congress, they had all fizzled.

That Tuesday in the hearing room, Zuckerberg adopted a more cooperative tone than in the past. When challenged on Facebook's long history of fighting regulations, he said he welcomed the "right regulation" in the United States, and he confirmed that Facebook would implement European privacy mandates introduced that year to protect all of Facebook's global users. "The expectations on internet companies and technology companies overall are growing," he said. "I think the real question is 'What is the right framework for this?' not 'Should there be one?'"

Zuckerberg talked a good game about the platform's commitment to security and data privacy, but the general consensus inside the company was that growth came first and that safety and security were an afterthought. Engineers were given engagement targets, and their bonuses and annual performance reviews were anchored to measurable results on how their products attracted more users or kept them on the site longer. "It's how people are incentivized on a day-to-day basis," one former employee recalled. On March 20, Sandy Parakilas, the operations manager who had warned about the Open Graph tool, noted in a *Washington Post* op-ed that in his sixteen months working at Facebook, he never saw "a single audit of a developer where the company inspected the developer's data storage." He believed the explanation for lax enforcement was simple: "Facebook didn't want to make the public aware of huge weaknesses in its data security."

Indeed, despite Zuckerberg's assurances to Congress, Facebook was waging a full-scale war against U.S. privacy regulations. Kaplan had built a formidable DC team, with fifty lobbyists, and was on track to spend $12.6 million that year, which made his operation the ninth-biggest corporate lobbying office in Washington. Facebook spent more on lobbying than oil giant Chevron or ExxonMobil and more than drug giant Pfizer or Roche. Kaplan had also turned Facebook into a powerful political force using the company's deep-pocketed PAC to fund political campaigns, doling out donations to Republicans and Democrats equally. It was important to keep alliances balanced, Kaplan had preached to his staff. In fact, Facebook's PAC had donated to the campaigns of more than half of all the lawmakers questioning Zuckerberg over the two days of testimonies.

Weeks after the hearing, Kaplan would meet privately with the top lobbyists of IBM, Google, and other tech giants at the Washington offices of their trade group, the Information Technology Industry Council. A California privacy bill was advancing that would be much stronger than the European Union's General Data Protection Regulation, a landmark law that would make it harder for Facebook to collect data and that would allow internet users to see what data was being collected. Kaplan would propose that the tech companies suggest a federal law on privacy with fewer restrictions than the California law, one that would also preempt state legislation. He was leading an industry fight for the most permissive regulations on privacy.

The hearing was "the most intense public scrutiny I've seen for a tech-related hearing since the Microsoft hearing," Sen. Orrin Hatch, the eighty-four-year-old Republican of Utah, said. It was about one hour into questioning, and several news outlets were carrying live blogs describing the lawmakers as tough and unrelenting and Zuckerberg as under siege.

But then the questioning took an unexpected turn.

"How do you sustain a business model in which users don't pay for your service?" Hatch asked. The senator didn't seem to understand the most basic details of how Facebook worked.

Zuckerberg paused and then smiled. "Senator, we run ads." The audience chuckled, and executives in Washington and Menlo Park exploded with laughter.

In the joint Senate Committee on Commerce and Senate Committee on the Judiciary hearing, the average age of its four leaders was seventy-five. The average age of all the members of the combined committees was not much younger. They were not the social media generation. Several lawmakers stumbled over the basic functions of Facebook and its other apps, Instagram and WhatsApp.

Zuckerberg was looking more comfortable. None of the questions forced him to deviate from his script. When he was asked, about two hours in, if he needed a break, he declined. "You can do a few more," he said with a small smile.

In Washington and Menlo Park, employees cheered. "Oh, he's feeling good!" one Washington staffer exclaimed.

Even younger congressional members made embarrassing errors. "If I'm emailing within WhatsApp, does that ever inform your advertisers?" asked Sen. Brian Schatz, forty-five, from Hawaii, earlier in the hearing. The app was for messaging, not for emailing, Zuckerberg explained, and all messages were encrypted. Four hours in, a lawmaker asked if Facebook listened in on voice calls to collect data for advertising. Others tripped up on jargon for how terms of service were used and how data was stored.

As some of the lawmakers exposed their deep knowledge gap in technology, the spotlight was shifting from Facebook's harmful business model to America's Luddite elected officials. "Lawmakers seem confused about what Facebook does—and how to fix it," a headline from Vox read. The hearings would become fodder for late-night television hosts Jimmy Kimmel and Stephen Colbert, who ran video reels on members' bad tech questions. Regulators tweeted that there needed to be new divisions within Congress and regulatory agencies to bolster tech expertise. In a stroke of enormous luck, the public shifted its ire toward Washington.

Zuckerberg returned to Silicon Valley at the end of the second day of hearings. Once his flight back to San Francisco was wheels up, staff in Washington celebrated at a wine bar in Georgetown. Shares of Facebook made their biggest daily gain in nearly two years, closing up 4.5 percent. Over ten hours of testimony and six hundred questions, Zuckerberg had added three billion dollars to his wealth.

$56b

2018

Chapter 9

Think Before You Share

The scope of the hearings had extended beyond the Cambridge Analytica scandal. Lawmakers fired off questions about the harmful addiction of the technology, the deceptive terms of service, and the election disinformation. They asked Facebook to share internal data about how it was working to protect users and whether it was in compliance with global privacy laws. And they brought up how, in the Southeast Asian nation of Myanmar, the spread of disinformation was playing out in the most visceral example of the platform's harms to date.

Genocide was taking place on Facebook in real time. On August 26, 2017, Sai Sitt Thway Aung, an infantryman from the Ninety-Ninth Light Division of the Burmese military, wrote a post for his five thousand Facebook followers: "One second, one minute, one hour feels like a world for people who are facing the danger of Muslim dogs." He was not the only soldier updating his Facebook page in the middle of a mass murder; human rights groups found dozens of other posts spouting similar anti-Rohingya propaganda, at times even including photographs of their advance along the dense jungle of Rakhine State. Nongovernmental organizations (NGOs) took screenshots, trying to document as much on the unfolding human rights crisis as they could. The official branches of the Burmese government denied responsibility for the violence and released false accounts of what was happening on the ground. The Facebook accounts of the soldiers were small flares of real information as to where, at least, the Burmese military was launching its attacks.

Facebook had thrown a lit match onto decades of simmering racial tensions in Myanmar and had then turned the other way when activists pointed to the smoke slowly choking the country. In March 2018, the United Nations' independent fact-finding mission in Myanmar told reporters that social media had played a "determining role" in the genocide. Facebook had "substantively contributed to the level of acrimony and dissension and conflict," Marzuki Darusman, chairman of the UN mission there, said. "Hate speech is certainly of course a part of that. As far as the Myanmar situation is concerned, social media is Facebook, and Facebook is social media," he said.

Human rights officials estimated that the genocide carried out by Burmese soldiers resulted in the killing of more than 24,000 Rohingya. Over the course of the next year, 700,000 Rohingya Muslims fled over the border to Bangladesh, seeking safety from Burmese troops in squalid refugee camps.

"What's happening in Myanmar is a terrible tragedy, and we need to do more," Zuckerberg had assured the U.S. lawmakers during the April hearings. The company would hire "dozens more" Burmese speakers and would work more closely with civil society groups to take down the accounts of local leaders who were spreading disinformation and calling for violence. (At the time, the platform had only five Burmese speakers to patrol content for Myanmar's eighteen million Facebook users. None of the native speakers was based in Myanmar.)

Five months later, Sandberg would sit before the Senate Intelligence Committee and call the situation in Myanmar "devastating." In the week before her hearing alone, the company took down fifty-eight pages and accounts in Myanmar, many of which posed as news organizations. "We're taking aggressive steps, and we know we need to do more."

In fact, the company had received repeated warnings for years that it needed to do more.

By March 2013, Matt Schissler had grown increasingly worried about the rumors circulating in Yangon. Friends who were Buddhist, the majority population of Myanmar, showed the American aid worker grainy cell phone photos of bodies they said were of Buddhist monks killed by Muslims. Others shared conspiracy theories about plots by the Rohingya, a long-resented Muslim minority.

"People wanted to tell me about all the ways in which Muslims were bad. Islamophobic stuff was suddenly coming up all the time, in every conversation," Schissler recalled. At six two, with close-cropped brown hair and blue eyes, he stood out as obviously American. "People had this idea that because of 9/11, I, as an American, would hate Muslims. They thought I would sympathize with them hating Muslims."

The neighbors Schissler had been getting to know around his urban neighborhood of Yangon were suddenly brandishing their cell phones to show him articles—some from reputable outlets like the BBC, others based on unnamed sources and dubious evidence—that claimed that Islamic State jihadists were en route to Myanmar. A journalist friend called to warn Schissler of a vague plot by Muslims to attack the country; he said he had seen a video showing the Muslims plotting to attack. Schissler could tell that the video was obviously edited, dubbed over in Burmese with threatening language. "He was a person who should have known better, and he was just falling for, believing, all this stuff," Schissler observed.

The spread of the doctored photos and videos coincided with Myanmar's introduction to cell phones. While the rest of the world went online and embraced mobile technology, Myanmar's military dictatorship had made it impossible for the average person to do either. But by the start of 2013, as military rule began to ease and as the government allowed foreign telecommunications operators into the nation, cell phone costs plummeted, and cheap used smartphones flooded the market. They came loaded with the internet and Facebook.

"Blue," the name used within the company for its flagship app, was their gateway online—the first, and often the only, app people in Myanmar used. The company's app was so popular that people there used the words *Facebook* and *internet* interchangeably. In the largest cities, where shops hawking mobile phones lined block after block of crowded streets, vendors helped users create Facebook accounts.

In a country where military rulers controlled newspapers and radio, Facebook felt like a bastion for individual expression. The Burmese were quick adopters of the technology. They shared family photos, egg curry recipes, opinions, memes, and folk stories. For those not yet online, a monthly magazine called *Facebook* was published that aggregated postings found on the social network.

The app was also a favorite of military figures and religious propagandists. Ashin Wirathu, a Buddhist monk whose anti-Muslim positions earned him the nickname "the Buddhist bin Laden," was quick to discover the platform's power. In the early 2000s, he had been arrested for helping incite an anti-Muslim riot. But with Facebook, his operation and ideas became professionalized. Wirathu had an office in Mandalay Bay, where his students ran several Facebook pages under his name that advocated for violence against the

Rohingya. They posted altered videos and photographs of corpses they claimed were of Burmese Buddhists who had been killed by Rohingya militants. They also stoked fears with rumors that the Rohingya were planning to attack civilian Buddhist populations in Mandalay and Yangon. None of the attacks materialized, but the rumors spread widely and contributed to panic and anger among the general population. Other monks and military generals were sharing similar posts that were then widely distributed by Burmese nationals. Dehumanizing photos and videos compared the Rohingya to vermin, insects, and rodents. People also posted videos meant to look science-based that asserted that Rohingya DNA was different from that of other people.

The ideas were not new. For years, state-sponsored radio and newsletters had made similar vitriolic claims against the Muslim minority. But Facebook enabled civilians to become the new bullhorns of hate. Individuals were passing along dangerous rhetoric they claimed had come from friends and family. Divorced from state media, which everyone knew were arms of propaganda for the government, the hate speech was more widely accepted and endorsed.

Schissler sent an email to donors and diplomats he knew from the NGO world in which he worked, suggesting that it would be worthwhile to look at the trajectory of Facebook's popularity in the region to see if it coincided with the change in information diet they were witnessing. He was worried about the role echo chambers played on the platform, he wrote, and thought it was worth studying their effect in Myanmar.

Nobody took up his call.

In August 2013, ahead of his thirtieth birthday and Facebook's tenth anniversary, Zuckerberg had tapped out a blog post on his iPhone to announce his vision for the next decade. The post was titled "Is Connectivity a Human Right?," and in it, he announced that Facebook, already the world's biggest communications network, with 1.15 billion users, was aiming to reach the next 5 billion customers.

It was his moonshot, a personal ambition that would put him in the company of tech visionaries like his mentors, Steve Jobs, who had revolutionized mobile computing in 2007 with the iPhone, and Bill Gates, who had transformed global philanthropy after revolutionizing personal computing. People close to Zuckerberg said he was frustrated with the negative press around his first major effort at philanthropy, a $100 million donation to the Newark, New Jersey, public school system in 2010 for education reforms. Critics panned the effort for doing little to significantly change the education system in the beleaguered city. Zuckerberg had begun to think more about his legacy and had shared with people close to him that he wanted to be remembered as an innovator and philanthropist in the mold of Gates, whom he publicly acknowledged as a personal hero in an interview at the *TechCrunch* Disrupt conference in 2013.

Internet connectivity was the great bridge to close the gap on global economic inequality, Zuckerberg wrote in the blog. Internet access led to stronger economies and higher gross domestic product; the vast majority of the 2.7 billion people with online access were from developed Western nations. In emerging markets, internet access was often restricted to male heads of households. "By bringing everyone online, we'll not only improve billions of lives, but we'll also improve our own as we benefit from the ideas and productivity they contribute to the world."

The proposition was hardly original. As David Kaye, former UN special rapporteur on freedom of expression, pointed out, "It was an idea that was being discussed on the world stage, by politicians and activists. The idea that people around the world should be brought online was part of the popular discourse." For years, the United Nations, Human Rights Watch, and Amnesty International had been advocating for internet companies to turn their attention to the developing world. But within their pleas was a note of caution that companies muscling into the new markets be cautious of local politics and media environments. Too much access to the internet too soon could be dangerous. "Everyone agreed that the internet was necessary to enjoy information and that universal internet access was an important goal," Kaye recalled. "But there was concern about whether the companies were doing the appropriate assessments about the countries and markets they were entering. Would a private company have the motivation to behave responsibly?"

In August 2013, Zuckerberg established Internet.org, a project with six global telecommunications partners aimed at bringing the whole world online. Internet.org struck business deals with cell carriers to offer a stripped-down internet service to developing nations. Facebook was preloaded and was compressed so it could be used even with slow and patchy internet connections. For remote areas with no cellular infrastructure, Zuckerberg created a laboratory for telecommunications projects like Aquila, an autonomous drone designed to beam down the internet to people below, or Catalina, which envisioned bird-size drones that could boost smartphone data speeds. Neither project made it beyond the testing phase.

Google had its own laboratory for broadband projects, which included hot-air balloons that beamed internet connections to rural areas of the world. The global race to acquire new internet customers was under way: Microsoft, LinkedIn, and Yahoo were also investing heavily in global expansion. Chinese companies like Weibo and WeChat were aggressively trying to expand beyond Asia, into Latin America and Africa. Zuckerberg was particularly focused on competing against Chinese companies head-on on their own turf. He had personally begun to lobby China's regulators and leaders, meeting with President Xi Jinping twice in 2015. The first to capture untapped global markets would be the best positioned for future financial growth.

"It was clear to everyone at Facebook that this was the thing Mark was most excited about, it had buzz," said a Facebook employee who worked on the initiative. In his weekly meetings with executives, Zuckerberg would regularly ask how new products in development would help with the "Next One Billion" project, and whether engineers were designing with the needs of the developing world in mind. "The message was clear that he wanted to get us there and get us there fast."

He wasn't thinking about the consequences of expanding so quickly, especially in nations that did not have democratic systems. As Facebook entered new nations, no one was charged with monitoring the rollouts with an eye toward the complex political and cultural dynamics within those countries. No one was considering how the platform might be abused in a nation like Myanmar, or asking if they had enough content moderators to review the hundreds of new languages in which Facebook users across the planet would be posting. The project didn't include an overseer role, which, as part of the policy and security staff, would have fallen under Sandberg's charge. It would have been a natural fit, given Sandberg's experience at the World Bank, but she acted more as a promotor and public advocate. "I can't recall anyone at the company directly questioning Mark or Sheryl about whether there were safeguards in place or raising something that would qualify as a concern or warning for how Facebook would integrate into non-American cultures," said one former Facebook employee who was closely involved with the Next One Billion project.

As it was, Facebook entered the markets, hired a few moderators to help review the content, and assumed the rest of the world would use the platform in much the same way it had been used in the United States and Europe. What happened in other languages was invisible to leaders in Menlo Park.

Zuckerberg, for his part, was encouraged by the early results. After Internet.org began rolling out in 2014, he touted how women in Zambia and India used the internet to support themselves and their families financially. He was excited about Facebook entering new markets like the Philippines, Sri Lanka, and Myanmar, and was undaunted by early critics. "Whenever any technology or innovation comes along and it changes the nature of something, there are always people who lament the change and wish to go back to the previous time," he conceded in an interview with *Time* magazine. "But, I mean, I think that it's so clearly positive for people in terms of their ability to stay connected to folks."

Lost in the excitement were the clear warning signs. On March 3, 2014, Matt Schissler was invited to join a call with Facebook on the subject of dangerous speech online. He had connected with a Harvard professor named Susan Benesch, who had published papers on hate speech and was communicating her concerns to members of Facebook's policy team. She asked Schissler to listen in on the call and give his perspective from Myanmar.

When he dialed in to the video link, he was introduced to a half-dozen Facebook employees and a handful of academics and independent researchers. Arturo Bejar, Facebook's head of engineering, was also on the call. Toward the end of the meeting, Schissler gave a stark recounting of how Facebook was hosting dangerous Islamophobia. He detailed the dehumanizing and disturbing language people were using in posts and the doctored photos and misinformation being spread widely.

The severity of what Schissler was describing didn't seem to register with the Facebook representatives. They seemed to equate the harmful content in Myanmar to cyberbullying: Facebook wanted to discourage people from bullying across the platform, they said, and they believed that the same set of tools they used to stop a high school senior from intimidating an incoming freshman could be used to stop Buddhist monks in Myanmar from spreading malicious conspiracy theories about Rohingya Muslims. "That was how Facebook thought about these problems. They wanted to figure out a framework and apply it to any problem, whether that was a classroom bully or a call for murder in Myanmar," said one academic who joined the call and who recalled that no one at Facebook seemed to probe Schissler for more information on the situation in Myanmar.

Schissler had spent nearly seven years in the region—first, along the border between Thailand and Myanmar and, later, within Myanmar itself. He had become conversant in Burmese and had studied the region's culture and history. What Schissler and other experts were seeing in Myanmar was far more dangerous than a one-off remark or an isolated Facebook post. Myanmar was consumed by a disinformation campaign against the Rohingya, and it was taking place on Facebook.

In the month following the call, a handful of Facebook employees started an informal working group to connect Facebook employees in Menlo Park with activists in Myanmar. The activists were told it would be a direct channel of communication, used to alert the company to any problems. Various members of Facebook's policy, legal, and communications teams floated in and out of the group, depending on the topics under discussion.

Just four months later, in the first week of July, the Myanmar activists had an opportunity to put the communication channel to the test as rumors began to spread on Facebook that a young Buddhist woman in Mandalay had been raped by Muslim men.

Within days, riots broke out across the country. Two people were killed, and fourteen were injured.

In the days leading up to the riots, NGO workers tried to warn the company in the private Facebook group, but they hadn't heard back from anyone. Now people were getting killed, and there was still no response.

On the third day of the riots, the Burmese government decided to shut down Facebook for the entire country. With the flick of a switch, the nation lost access to the platform. Schissler reached out to Facebook through a contact to ask if they knew about the problem, and he heard back almost immediately. "When it came to answering our messages about the riots, Facebook said nothing. When it came to the internet being shut down, and people losing Facebook, suddenly, they are returning messages right away," recalled another activist in the group. "It showed where their priorities are."

The week before the riots, Schissler had taken to the Facebook group to let the company know that an elderly man was being maligned on the site. The man worked at a charity and did not have a Facebook account. He had been photographed delivering rice and other foods to camps for Rohingya who had been displaced. His image was then shared to Facebook, where people accused him of "helping the enemy" and threatened violent attacks.

Schissler had reported the photograph through Facebook's automated systems, only to be told that it did not contain any images that were hateful or threatening. When he explained that it wasn't the photo but the posts and comments underneath it that were problematic, he received no answer. He then asked Facebook staff to take action. Weeks later, he was told they could do nothing unless the man in the photograph reported the image to Facebook himself—an impossibility given that he didn't have a Facebook account.

It wouldn't be the last time Schissler detected a Kafkaesque bureaucracy at play in the company's response to the situation in Myanmar. A few months after the riots and the internet shutdown, he discovered why the platform might be so slow to respond to problems on the ground, or even to read the comments under posts. A member of Facebook's PR team had solicited advice from the group on a journalist's query regarding how Facebook handled content moderation for a country like Myanmar, in which more than one hundred languages are spoken. Many in the group had been asking the same thing, but had never been given answers either on how many content moderators Facebook had or on how many languages they spoke.

Whatever Schissler and the others had assumed, they were stunned when a Facebook employee in the chat typed, "There is one Burmese-speaking rep on community operations," and named the manager of the team.

The manager then joined the conversation, confirming, "We have one contractor in the Dublin office who supports the Burmese community."

Everything was beginning to make sense. There was only one moderator tasked with monitoring the deluge of content being generated in Myanmar, and that person spoke only Burmese. The repeated complaints of harassment and hate speech had gone unanswered because the company was hardly paying attention. The sheer amount of information being shared in Myanmar was tremendous, but Facebook hadn't expanded its staff to keep up.

"There are maybe one hundred languages spoken, and more that are dialects, but Facebook thought it was enough to have Burmese," said one Myanmar-based activist who had seen the post go up in the Facebook group. "It would be like them saying, well, we have one German speaker, so we can monitor all of Europe."

After the July riots, Facebook's communications team released public statements promising to work harder to protect the safety of users in Myanmar. Within their Facebook group, employees asked Schissler and other activists to help translate their community standards into Burmese. A Burmese NGO, Phandeeyar, partnered with Facebook on the translation. It also created a public service campaign around digital "sticker packs," with messages users could add to posts to prevent the spread of harmful speech. The stickers featured cartoon figures of a young girl, a grandmother, or a crying child with a button nose and rosy cheeks. They included the messages "Think before you share" and "Don't be the cause of violence." The stickers, Facebook said, helped moderators spot harmful posts and more quickly remove them.

But Schissler and others realized the stickers were having an unintended consequence: Facebook's algorithms counted them as one more way people were enjoying a post. Instead of diminishing the number of people who saw a piece of hate speech, the stickers had the opposite effect of making the posts more popular.

Schissler was growing frustrated with the company's lack of response and cut back his participation in the online group. He would make one final effort, however, to get through to Facebook. In March 2015, he combined a personal trip to California with a visit to MPK. In a small conference room where roughly a dozen Facebook employees had gathered, with others joining by video-conference, he shared a PowerPoint presentation that documented the seriousness of what was happening in Myanmar: hate speech on Facebook was leading to real-world violence in the country, and it was getting people killed.

Later, he met with a smaller group of employees who were interested in continuing the discussion. Toward the end of the meeting, one of the Facebook employees turned to him with a question. With a furrowed brow, she asked for Schissler to predict what would happen in Myanmar in the coming months, or years. Was it possible, she asked, that people were blowing the violence out of proportion? She had heard the word *genocide* used, but did Schissler think that could really happen in Myanmar?

"Absolutely," he answered. If Myanmar continued on its current path, and the anti-Muslim hate speech grew unabated, a genocide was possible. No one followed up on the question. After Schissler left, two Facebook employees lingered in the hallway outside the conference room. "He can't be serious," one said loudly. "That's just not possible."

The root of the disinformation problem, of course, lay in the technology. Facebook was designed to throw gas on the fire of any speech that invoked an emotion, even if it was hateful speech— its algorithms favored sensationalism. Whether a user clicked on a link because they were curious, horrified, or engaged was immaterial; the system saw that the post was being widely read, and it promoted it more widely across users' Facebook pages. The situation in Myanmar was a deadly experiment in what could happen when the internet landed in a country where a social network became the primary, and most widely trusted, source of news.

Facebook was well aware of the platform's ability to manipulate people's emotions, as the rest of the world learned in early June 2014, when news of an experiment the company had secretly conducted was made public. The experiment laid bare both Facebook's power to reach deep into the psyche of its users and its willingness to test the boundaries of that power without users' knowledge.

"Emotional states can be transferred to others via emotional contagion, leading people to experience the same emotions without their awareness," Facebook data scientists wrote in a research paper published in the *Proceedings of the National Academy of Sciences*. They described how, over the course of one week in 2012, they had tampered with what almost 700,000 Facebook users saw when they logged on to the platform.

In the experiment, some Facebook users were shown content that was overwhelmingly "happy," while others were shown content that was overwhelming "sad." A happy post could be the birth of a baby panda at the zoo. A sad post could be an angry editorial about immigration. The results were dramatic. Viewing negative posts prompted users to express negative attitudes in their own posts. And the more positive content a user saw, the more likely that user would spread upbeat content.

As far as the study was concerned, this meant that Facebook had demonstrated that emotional contagion could be achieved without "direct interaction between people" (because the unwitting subjects were seeing only each other's News Feeds). It was a dramatic finding.

Once tech and science reporters discovered the paper, it was only a matter of time before it made headlines in the mainstream press. "Facebook Treats You like a Lab Rat," blared one CNN headline. The mood experiment confirmed critics' assertions of the company's power to influence ideas and stir outrage; Facebook was operating behind a curtain, like the Wizard of Oz. When press reports and consumer groups put pressure on the company to explain itself, Sandberg instead issued a statement in which she apologized for the company's clumsy PR. "This was part of ongoing research companies do to test different products, and that was what it was; it was poorly communicated. And for that communication we apologize. We never meant to upset you."

But Facebook's engineers, who were not aware of the experiment before the paper was published, recognized that it revealed something more nefarious about the News Feed's ability to influence people. Over the years, the platform's algorithms had gotten more sophisticated at identifying the material that appealed most to individual users and were prioritizing it at the top of their feeds. The News Feed operated like a finely tuned dial, sensitive to that photograph a user lingered on longest, or the article they spent the most time reading. Once it had established that the user was more likely to view a certain type of content, it fed them as much of it as possible.

The engineers saw a problem developing and repeatedly tried to raise the alarm with their managers. In a series of posts on Workplace groups, they complained about the clickbait sites' routinely being promoted to the top of the News Feed, and they asked if the algorithm needed to be reevaluated. One engineer posted a series of examples of false or salacious stories that were going viral, including posts claiming that President Barack Obama had secretly fathered a love child and that a cabal of global elites was hoarding a magical serum for eternal youth. "Is this really the first thing people should be seeing in their feeds?" he asked.

In the fall, several employees confronted Cox and asked him to take their concerns to Zuckerberg. They insisted that he explain to the CEO that something in the algorithm was deeply broken. "Chris told us he sympathized. He saw the problem, too," one engineer recalled. "We compared what the News Feed was serving up to junk food. It was sugary junk, and we knew it, and it wasn't healthy, for anyone involved, for us to keep promoting it."

And yet Cox was torn. Facebook's users were spending more time on the platform than ever before, and that meant that the systems were working. Sugary junk was addictive, and everyone knew it. Cox held a long view of the News Feed's merits.

A few weeks later, the engineer followed up with Cox. The company was taking the issue seriously, Cox assured him. There were changes currently under development that would impact the News Feed.

Those changes would take more than six months to unfold and

would have little material effect. In June 2015, Facebook announced a change to the News Feed that emphasized "time spent" as the most important metric of activity on Facebook. The assumption was that users quickly lost interest in clickbait content, and by prioritizing the duration of interest in content, more legitimate news stories would surface higher in the rankings. The company also began a process of down-ranking sites they believed were "clickbait farms," or websites that churned out meaningless junk content with no goal other than getting a user to click on it.

Users still clicked on, and spent a significant portion of time reading, junk content, and clickbait sites tweaked their headlines and approaches to avoid being down-ranked. And sensationalist posts continued to rise to the top of the News Feed.

The harms were baked into the design. As Dipayan Ghosh, a former Facebook privacy expert, noted, "We have set ethical red lines in society, but when you have a machine that prioritizes engagement, it will always be incentivized to cross those lines."

In late September 2018, Matthew Smith, the CEO of Fortify Rights, a human rights organization based in Southeast Asia, began to work with human rights groups to build a case strong enough for the International Criminal Court, at The Hague, proving that Burmese soldiers had violated international laws and perpetuated a genocide against the Rohingya. They needed evidence, and they believed Facebook had data that would help. The platform held detailed information on all its user accounts; even when posts were deleted, Facebook kept a record of everything a person had ever written, and every image uploaded. Its mobile app also had access to location information. Most Burmese soldiers had Facebook on their phones, so the company would have records of the locations of army units' soldiers to match with attacks on Rohingya villages.

Officials at Facebook had recently removed thousands of Facebook accounts and pages secretly run by the Burmese military that were flooded with hate speech, disinformation, and racist tropes. The posts included photos of bombings or corpses the user had identified as images of innocent people killed in attacks by the Rohingya, but the photos were often actually stills from movies or stolen newspaper stock images from Iraq or Afghanistan. Facebook took down the posts following a front-page story that ran in the October 16 issue of the *New York Times*.

If Smith and other human rights workers could get their hands on the deleted posts, they could build a stronger case documenting how Myanmar's military had both carried out a genocide against the Rohingya and manipulated the public into supporting their military onslaught. Smith approached members of Facebook's policy and legal teams and asked them for their cooperation and to give the human rights groups evidence to prosecute. "Facebook had a lot of actionable information that could be used in prosecution for crimes," Smith said. "They had information that could connect soldiers to where massacres took place. I told them that that could be used by prosecutors looking to bring people to justice at the International Criminal Court."

The lawyers at Facebook refused the request, claiming that handing over data could be a violation of privacy terms. Then they got bogged down in legal technicalities. Soldiers could claim they were carrying out orders from military leaders and sue Facebook for exposing them. Facebook would cooperate, Smith was told, only if the United Nations created a mechanism to investigate human rights crimes. When Smith pointed out that the United Nations had such a system in place, known as the Independent Investigative Mechanism for Myanmar, the Facebook representative looked at him with surprise and asked him to explain. "I was shocked. We were meeting with Facebook to talk about international justice in

Myanmar, and they didn't know the basic frameworks created by the UN," Smith said.

Facebook's lawyers also said the company didn't have an internal process in place to find the harmful content Smith was requesting. The answer was misleading. For years, the company had worked with U.S. law enforcement to build cases against child predators.

"Facebook had the chance to do the right thing again and again, but they didn't. Not in Myanmar," said Smith. "It was a decision, and they chose not to help."

$56b

2018

Chapter 10

The Wartime Leader

The conference room where the M-Team met that July 2018 morning was called "the Son of Ping and Pong," a tongue-in-cheek reference to a conference room in the original office that was located near the Ping-Pong tables. The group, which comprised more than twenty top executives and managers, strained to appear upbeat. They had endured eighteen months of one bad news cycle after another. They had been forced to defend Facebook to their friends, family, and angry employees. Most of them had had little to do with the controversies over election disinformation and Cambridge Analytica, and several of them were struggling with their private frustrations with Zuckerberg and Sandberg.

Held two or three times a year, M-Team gatherings were times of bonding. Zuckerberg would kick things off with a boozy dinner catered by one of his favorite Palo Alto restaurants. At the daytime sessions, executives gave exuberant presentations, with charts showing hockey stick growth in revenue and users. They showcased audacious plans for products in artificial intelligence, virtual and augmented reality, and blockchain currency. The group hashed out the biggest problems facing Facebook and discussed ways to beat back competition.

The forty or so executives had gotten to know one another well over the years, working through Facebook's painful transition to mobile, its rocky IPO, and its race for the first billion and then the second billion users. They often veered off the topic of business to share stories of marriage, divorce, new babies, or unexpected deaths in the family. Zuckerberg would cap off the two-day meeting with a rallying motivational speech before sending them back to lead their fiefdoms of engineering, sales, products, advertising, policy, and communications.

Zuckerberg held court, as always, at the center seat of a large U-shaped table. Boz gave a rousing presentation on plans for a video-calling device called "Portal," Facebook's attempt to compete with home devices like Amazon's Echo and Google's Home Hub. Another executive presented an update for Oculus, the company's VR headset. Zuckerberg's face brightened. The product unveilings were his favorite part of the M-Team meetings. They showed that rather than resting on his laurels, Zuckerberg was still pushing his company to innovate.

Sandberg spoke next. The company was getting back on track, she told the group. The scandals of the past year were behind them; Zuckerberg's performance at the congressional hearings had been a triumph. She reminded everyone of the company's mission statement, which Facebook had officially debuted the previous year: "Give people the power to build community and bring the world closer together."

"We're putting the wheels back on the bus," she concluded.

But the moment of optimism was fleeting. During their presentations, several executives expressed concern for employees. Internal surveys showed that workers felt demoralized by Facebook's role in election interference and the Cambridge Analytica scandal. Turnover was high, and recruiters struggled to attract new engineers. Recent college graduates said the idea of working at Facebook, once the hottest place to work in Silicon Valley, had lost its appeal. In the intense competition for engineering talent, employee satisfaction was a priority.

Zuckerberg listened to his deputies with his famously unnerving stare. In one-on-one conversations, he could hold an unwavering gaze for minutes. This created painful stretches of dead air. Long-time executives charitably described the quirk as Zuckerberg's mind consuming and processing information like a computer.

When his turn came, Zuckerberg drew a long pause. Then, in a surprising change of topic, he began to talk about his philosophy of leadership. Facebook had evolved, he said. It was now a sprawling enterprise with more than thirty-five thousand employees, seventy offices in thirty-five countries, more than a dozen products, and fifteen multibillion-dollar data centers around the world. Up until this point, the social network had faced obstacles with competition but had enjoyed a clear runway of growth and goodwill from the public, buoyed by the techno-optimism of the past two decades. But that period was over. Consumers, lawmakers, and public advocates had turned on Facebook, blaming it for addicting society to their smartphones and for poisoning public discourse. The business had become the poster child for irresponsible, at-all-costs growth. It faced multiple investigations by regulators around the world. Shareholders, users, and consumer activists were pursuing lawsuits in court. "Up until now, I've been a peacetime leader," Zuckerberg declared. "That's going to change."

He had been influenced by a book called *What You Do Is Who You Are*, by Ben Horowitz, one half of Andreessen Horowitz, the VC firm that invested early in Facebook. (Marc Andreessen, Zuckerberg's friend and a Facebook board member, was the other half.) In his book, Horowitz argues that at various stages of development, tech companies demand two kinds of CEOs: wartime and peacetime leaders. In periods of peace, he writes, a company can focus on expanding and reinforcing its strengths. In times of war, the threats are existential, and the company has to hunker down to fight for survival. Google's CEO Eric Schmidt had led the search engine's growth into a global powerhouse during a period of peace and little competition. But Larry Page had taken over as CEO as the company saw new threats in social media and mobile technology. "Peacetime CEO works to minimize conflict. Wartime CEO heightens the contradictions," Horowitz writes. "Peacetime CEO strives for broad-based buy-in. Wartime CEO neither indulges consensus-building nor tolerates disagreements."

The idea was widely embraced by tech leaders, but it wasn't new. In public speeches and interviews, Horowitz acknowledged that he had been inspired by *Only the Paranoid Survive*, written by Intel's famously brutal former CEO Andy Grove, who had steered the company from the production of memory chips to processors. Grove was notorious in Silicon Valley for bullying challengers and opponents as part of his philosophy: never relax, because a new rival is coming just around the corner.

(In a blog post, Horowitz conceded that he had also taken some inspiration from the movie *The Godfather*, quoting a scene where consigliere Tom Hagen asks the Corleone family don, "Mike, why am I out?" and Michael Corleone replies, "You're not a wartime consigliere." The reference to *The Godfather*'s Mob boss struck some Facebook employees as apt. "It shouldn't surprise anyone that Mark would want to think of himself as modeling his behavior after Andy when in fact he was taking cues from Hollywood's most notorious gangster," said one M-Team member. "You won't find anyone more ruthless in business than Mark.")

From that day forward, Zuckerberg continued, he was taking on the role of wartime CEO. He would assume more direct control over all aspects of the business. He could no longer sequester himself to focus only on new products. More decisions would fall to him.

The M-Team was silent. Sandberg and a small group of executives who knew in advance what Zuckerberg had prepared to say that day nodded approvingly. But others could only keep their faces carefully neutral as they tried to assess what Zuckerberg's words meant for their departments and personal advancement. Technically, as the founder and stakeholder with a majority of voting shares since Facebook's earliest days, Zuckerberg had always been in control of the company. But he had liberally delegated many parts of the business. Executives in charge of arms like WhatsApp and Instagram could operate with a fair amount of independence, and powerful heads of engineering within Facebook's central product had cultivated their own, loyal teams. "Zuckerberg hired and promoted managers who had big dreams for themselves. For years, he had kept them happy because they knew they had their relative autonomy," noted one member of the M-Team who attended the meeting. "Now he seemed to be saying that was about to change."

Zuckerberg's affinity for the first Roman emperor, Caesar Augustus, whom he had studied at both Exeter and Harvard, was well known. Up to this point, Zuckerberg's governance had been analogous to that of the Roman Republic. There was a clear leader, but his power was backed by a Senate of sorts that deliberated over big decisions and policies. What he was outlining now looked more like Rome's move from a republic to an empire. Zuckerberg was establishing rule by one.

Predictably, Boz was one of the first to weigh in with an endorsement of the declaration. It was the best path forward for the company to put Zuckerberg at the center of all decision making, he said.

Zuckerberg's proclamation came on the heels of a major reorganization at the company. Two months earlier, he had overhauled the leadership ranks by reshuffling the duties of his deputies. Chris Cox, head of product, would oversee all of the company's apps, including Blue, WhatsApp, Facebook Messenger, and Instagram. The chief technology officer, Mike Schroepfer, was put in charge of all emerging products, such as virtual reality and blockchain. Javier Olivan, the head of the growth team, added security, ads, and services to his portfolio. Sandberg remained chief operating officer, but it was less clear if she was still, in effect, Zuckerberg's second-in-command. She hadn't been mentioned in the announcement. Now, with Zuckerberg asserting himself even further, it seemed he envisioned himself as number one—and then there was everybody else.

Discontented employees had been departing all year. Jan Koum, the cofounder of WhatsApp, announced his departure in April; he later explained that Zuckerberg had broken his promise to keep WhatsApp separate from the Facebook app and to protect privacy. Kevin Systrom, the cofounder of Instagram, left in September for similar reasons: Zuckerberg had reneged on promises. And of course, Stamos had officially given up his duties in August but, in fact, had given his resignation months earlier.

At age thirty-four, roughly fourteen years after founding Facebook, Zuckerberg was doubling down on the insignia he had typed into the bottom of the Facebook website in 2004: "A Mark Zuckerberg Production."

"From day one, it was his company," an executive at the meeting said. But "he was happy to let other people do the stuff he wasn't interested in. That all changed."

Zuckerberg may have changed his mind about delegating responsibility, but he had always exhibited a ruthless streak, especially when it came to competition. Even with the acquisition of Instagram in 2012, he had continued to fret over a new crop of apps that could draw users away from Facebook and put his company behind in the race for mobile supremacy. Then Javier Olivan, vice president of growth, told him about Onavo, an Israeli mobile analytics start-up. The company operated a series of consumer-facing apps, including a virtual private network for mobile phone users, but of greater interest to Olivan was Onavo's technology, which enabled mobile publishers to chart how well not only their own apps were performing but also those of their competitors. Onavo's analytics tracked activity such as the number of times a user visited other mobile apps, how long they used them, and what kinds of functionality were most popular in those apps. It was "really cool for identifying acquisition targets," Olivan wrote in an email. Zuckerberg was intrigued, and beginning in early summer, Olivan began to visit Tel Aviv to meet with the founders of the company to learn more about the capabilities of the service and how much competitive insight Onavo truly offered. After receiving Olivan's reports, Zuckerberg decided the data was not only valuable to Facebook but also potentially helpful to rivals.

On October 13, 2013, Onavo announced that it had been acquired by Facebook—along with the company's proprietary technology and data, which created in-depth reports on how people all over the world were using the internet. Facebook would be absorbing Onavo's thirty-odd employees in Israel and its founders, Guy Rosen and Roi Tiger.

Days after news of the acquisition—which carried an estimated price tag of $115 million—hit, Rosen, Tiger, and other Onavo executives met with their counterparts at Facebook to discuss the deal and to celebrate. They were peppered with questions they were told came from Zuckerberg: Could they study which applications were used in India versus Europe? Could they create reports on which new apps were gaining traction fastest, week after week? Could they tell Zuckerberg what apps people were using over Facebook, especially to message one another? It was all possible, the Onavo executives assured the Facebook team. "Onavo was Mark's new toy, and he was excited by it," recalled one member of the Onavo team. "He was looking at the data himself every day."

Zuckerberg was particularly focused on data about the messaging app WhatsApp, which had surpassed Facebook Messenger to become number one in the world in overall daily volume of messages sent—an estimated 12.2 billion a day, compared to Facebook's 11.7 billion. The Onavo data also showed that users spent longer periods on WhatsApp and that it was more popular across all age groups than Messenger. Zuckerberg feared that WhatsApp, a company founded by two former Yahoo engineers in 2009 that was popular around the world for its commitment to privacy, could easily adopt features that would make it more like a social network. He viewed multiple threats: it could independently challenge Facebook and its Messenger service or be bought by a competitor like Google to create a viable social networking rival. Four months later, Facebook announced its plan to acquire WhatsApp for $19 billion.

In 2018, Zuckerberg's interest continued to lie in the product side of the business. Issues such as Facebook's broken reputation and the mounting complications in Washington were simply problems of perception and could be finessed. Six months after Facebook's mishandling of private user data became public through the reporting on Cambridge Analytica, Zuckerberg was still in denial that there were broader issues with how much data Facebook collected. Nearly two years after Russian-backed hackers meddling in the 2016 elections used polarizing language and images to divide Americans against each other, he was still recalcitrant with regard to drawing the line on what constituted hate speech. It was clear that Zuckerberg would not change Facebook's platform or its core products. Instead, he wanted his COO to change people's minds.

Sandberg entered Facebook's DC office on September 4 and beelined into a conference room. Workers had covered the glass walls with contact paper. A security guard stood outside the closed door. Kaplan, his lobbyists, and a group of WilmerHale litigators greeted her inside.

Kaplan had moved his Washington staff into a new 170,000-square-foot open-concept office in the trendy and vibrant Penn Quarter, near the Capitol. The Washington office, which also housed sales and security employees, was designed to replicate MPK's open floor plans, with polished concrete floors, large modern art installations, and open ceilings exposing piping and electrical cords.

Sandberg normally gave a speech to the staff when she visited the DC office. Her tours through Washington were a big affair; she'd set up a war room to prepare for meetings with lawmakers and for staff to help write thank-you cards for every official. But on this visit, she dispensed with a staff meeting to prepare to testify before the Senate Intelligence Committee, which was overseeing an investigation into Russian disinformation campaigns on Facebook and other social media platforms.

The stakes couldn't have been higher. In the aftermath of Zuckerberg's reprimand for her handling of the blowback from the Cambridge Analytica and 2016 election scandals, Sandberg's private efforts with legislators had not been as effective as she had hoped. The hearing before Congress was a chance to win over not just the lawmakers but also journalists as she defended Facebook from charges of complicity in election interference.

That summer, following Zuckerberg's wartime declaration, Sandberg had begun an intense behind-the-scenes campaign of her own. If Zuckerberg was conducting a full PR assault with on-the-record interviews with senior journalists, she would work backroom channels to persuade powerful players in Washington to believe him. In the lead-up to the hearing, Sandberg insisted that she would only testify alongside the CEOs of Twitter and Google. She wanted to spread the blame across the tech industry, even though Facebook was the most embattled among the internet giants. Twitter CEO Jack Dorsey accepted the invitation, but Google's CEO, Sundar Pichai, refused to testify. It was the perfect press diversion. Ahead of the hearing, journalists ran articles about the company's absence, accompanied by photographs of an empty chair and a nameplate reading "Google." In background briefings with journalists, angry lawmakers pointed to Pichai's decision as hubris and made a point of thanking Sandberg and Dorsey for flying into Washington.

On the day of the hearing, Dorsey entered the hearing first, his beard unkempt and his shirt unbuttoned at the neck. Sandberg followed, in a conservative black suit. Where Dorsey had strolled into the congressional hearing with just a handful of aides and assistants, unsure of where to sit, Sandberg had entered with more than double his staff and immediately began shaking hands with members of Congress. When she took her seat, she opened her black plastic binder and placed a white sheet of paper with a list of the lawmaker's names on the desk. In big capital letters at the top of the page were the words SLOW, PAUSE, DETERMINED.

Sandberg left the hearing believing she had sailed through her testimony. In contrast to her smooth, practiced delivery, Dorsey gave rambling and unscripted answers. But in the press reports published that afternoon, Dorsey received rave reviews for his candor, while Sandberg's calculated responses earned poor marks.

One exchange in particular highlighted the differing styles of the two executives. When Republican senator Tom Cotton of Arkansas asked, "Would your companies ever consider taking these kinds of actions that privilege a hostile foreign power over the United States?" Sandberg waffled. "I'm not familiar with the specifics of this at all, but based on how you're asking the question, I don't believe so," she said.

Dorsey's answer was two words: "Also no."

Sandberg's over-rehearsed style of communication was familiar to that of government officials who had interacted with her throughout the years. A meeting in October 2010 in which she met with FTC chairman Jonathan Leibowitz to try to quell the privacy investigation into profile settings and the Open Graph program had been particularly memorable.

A relaxed and confident Sandberg began the meeting with a claim that Facebook had given users more control over their data than any other internet company and that the company's biggest regret was not communicating clearly how its privacy policy worked. The FTC officials immediately challenged her, with Leibowitz noting that, on a personal level, he had watched his middle school–age daughter struggle with privacy settings on Facebook, which had automatically made it easier for strangers to find users like her. "I'm seeing it at home," he said.

"That's so great," Sandberg responded. She went on to describe the social network as "empowering" for young users. She seemed to be hearing only what she wanted to hear.

"This is a serious issue, and we're taking our investigation seriously," Leibowitz retorted.

Sandberg then attempted to divert the conversation to how critics were wrong about the company's privacy abuses. She wasn't answering the commission's questions related to data privacy, but was instead speaking at a broad level about Facebook's contributions to the economy.

"It was like she thought she was on a campaign or something, not a meeting about a federal investigation," David Vladeck, the head of the consumer protection bureau overseeing the investigation, said. "We were surprised that someone as sophisticated as Sheryl Sandberg could be as tone-deaf as she was."

Some had a less charitable interpretation. "Arrogance is her weakness, her blind spot. She believes there is no person she can't charm or convince," a former Facebook employee observed.

Eight years later, she still wasn't reading the room. Following the hearing, one senator quipped that Sandberg gave the perfect, empty answers Congress had come to expect from Facebook. "It was a good performance, but we all saw it for what it was, a performance."

Sandberg wasn't the only Facebook executive to make an appearance at a high-profile hearing that month. Just three weeks after her DC visit, the nation tuned into the fraught public proceedings surrounding the confirmation of Judge Brett Kavanaugh to the Supreme Court. A high school peer, Christine Blasey Ford, and other women had accused Kavanaugh of sexual assault and harassment. But only Blasey Ford was allowed to testify. It was the height of the MeToo movement, and Kavanaugh's confirmation hearing had become a referendum on systemic sexual harassment and the abuse of women across business and politics.

An estimated twenty million people watched Kavanaugh's hearing on the sexual assault allegation, which began the morning of September 27. As he delivered his opening remarks, the cameras framed him at the witness table with his wife, Ashley Estes Kavanaugh, in view behind his left shoulder. Seated just one row back from her was Joel Kaplan.

Kaplan and Kavanaugh were close friends, arguably best friends. Their families were intertwined. The men, both in their late forties, had cut their teeth in Washington during the George W. Bush years. They were members of the Federalist Society and part of a tight-knit cohort of Ivy League conservative men in Washington of the same age who were involved in the Bush campaign or part of the Bush world. Kaplan saw his appearance at the hearing as a moment to support his good friend at the most trying time of his life. His wife, Laura Cox Kaplan, sat by Ashley Kavanaugh's side.

Journalists immediately recognized Kaplan and called communications staff at Facebook for confirmation. The PR staff fumbled; they had not known he would be attending the hearing and initially told journalists he had taken the day off. When reporters tweeted that Kaplan was in DC on his own time, some Facebook employees checked their work calendars and discovered that Kaplan was not scheduled for vacation that day; they complained on Workplace groups that he appeared to be at the hearing in an official capacity. The employees were later informed by their managers that a mistake had been made and that Kaplan had requested the time off—the system just hadn't been updated. But anger at Kaplan continued to boil over. One employee pointed out that it was implausible that the head of government relations did not know he would appear on camera during the hearing. "His seat choice was intentional, knowing full well that journalists would identify every public figure appearing behind Kavanaugh. He knew that this would cause outrage internally, but he knew that he couldn't get fired for it. This was a protest against our culture, and a slap in the face to his fellow employees," he wrote on the Workplace message board. "Yes, Joel, we see you," the employee added.

Employees complained to their managers and directly to Sandberg and Zuckerberg. Kaplan was already unpopular among some staff for his politics in general, and specifically for his guidance on policies related to speech and Trump. The swirl of MeToo and the political rancor within Facebook and across the nation all collided in his decision to appear at Kavanaugh's hearing.

Kaplan decided he had to address the staff regarding his decision. "I want to apologize," he wrote in an internal note. "I recognize this moment is a deeply painful one—internally and externally." He regretted catching employees off guard with his surprise appearance, but he did not regret attending the hearing. He dug in on his loyalty to his friend.

Sandberg wrote in a Workplace post that she had spoken with

Kaplan and told him that it had been a mistake to attend the hearing, given his role in the company. "We support people's right to do what they want in their personal time, but this was by no means a straightforward case."

When Kaplan's appearance at the hearing was raised in the Q&A later that week, Zuckerberg said the VP of public policy hadn't broken any rules. He added that he probably wouldn't have done the same, however.

Kaplan did not seem to have garnered any punishment or scolding from Zuckerberg and Sandberg for his behavior. ("You don't get fired for stupidity, you get fired for being disloyal," a Facebook employee observed.)

To many, it was a reminder that the interests of protecting the business outweighed any consideration of the company's employees. Kaplan played a crucial role in the royal court of Facebook, and he played it well; he would be protected at all costs.

It had always been the case that conservative voices had a place at Facebook. Sandberg and Zuckerberg defended longtime board member and early investor Peter Thiel, Trump's emissary for tech types, when employees and shareholders called for his removal from the board for his support of the president. Thiel's hypocrisy on speech rankled employees. He extolled free expression but had also bankrolled Hulk Hogan's privacy suit against Gawker, a website that had posted portions of a sex tape featuring the professional wrestler. Gawker had outed the billionaire investor as a backer of the suit.

Some of the female employees expressed particular disappointment in Sandberg, whose "Lean In" mantra was supposed to empower women in the workplace. The figures on diversity at Facebook were only inching forward, with women filling 30 percent of leadership positions in 2018, up from 23 percent in 2014. Only 4 percent of employees were Black. Kaplan symbolized a culture of privilege that had been on trial across the nation with MeToo. Many employees saw Zuckerberg's and Sandberg's support of him as their complicity in that culture.

Days after the Q&A, Kaplan threw a party at his multimillion-dollar home in Chevy Chase, Maryland, to celebrate Kavanaugh's confirmation.

In July, Zuckerberg had held his first extended interview in years, with Kara Swisher, an editor of the tech news site Recode. His communications team cranked up the air-conditioning in the Aquarium to icebox mode for the interview. Swisher was a veteran journalist with a must-listen-to podcast that featured C-suite leaders of Silicon Valley. Her acerbic wit and tough questions drew a large and influential following. Many, Zuckerberg among them, found her intimidating.

Eight years earlier, Swisher had interviewed him onstage at the tech conference D: All Things Digital. At the time, Facebook was under scrutiny for changes to its privacy settings. As Swisher grilled him about the controversy, the then-twenty-six-year-old froze and spiraled into a flop sweat. Streams of perspiration rolled off his forehead and cheeks; his hair was plastered to his temples. Zuckerberg's discomfort was so apparent that Swisher paused and suggested he remove his hoodie. It was a humiliating appearance. His PR team later insisted that Zuckerberg simply ran hot and that the hoodie was too thick for the blazing stage lights.

In the intervening years, Zuckerberg had spoken at developer conferences and given occasional speeches, but he generally left interviews with the press to Sandberg. And yet his communications team was telling him now that the public needed to hear from him directly. Facebook needed to show that it was changing. He had declared his wartime command to his executives. He now needed to show the world that he was in control and steering Facebook back toward stability.

A new public relations executive, Rachel Whetstone, insisted Zuckerberg appear on Swisher's podcast. Senior members of the communications team prepped him on a range of issues Swisher would likely cover: Russian election interference, Cambridge Analytica, and privacy policies. The most recent controversy was over hate speech and misinformation.

For months, individuals and public interest groups had hammered Facebook and other social media sites for hosting the far-right talk show host Alex Jones and his conspiracy-laden site, Infowars. Jones attracted millions of followers on Facebook, Twitter, and YouTube with outrageous and demonstrably false claims and incendiary accusations against liberal political figures. His most harmful assertion was that a gunman's 2012 rampage at Sandy Hook Elementary School, in Newtown, Connecticut, which killed twenty-six people, including twenty first-grade children, was a hoax. Jones's audience believed him. The parents of one child killed during the attack had to move from their home because of death threats from Jones's followers.

Jones had accumulated more than a million followers on Facebook, and his shock jock–style tropes were primed for Facebook's algorithms. Even when people disagreed with Jones and left outraged comments on his stories or shared them in disgust, they were helping catapult his content to the top of News Feeds. Pressure was building for Facebook to ban Infowars and Jones. His posts violated Facebook's rules against hate speech and harmful content. But throughout 2018, Facebook refused to remove his accounts.

Predictably, less than halfway into the interview, Swisher pressed Zuckerberg on the topic. "Make a case," she said, for why he still allowed Jones's content to reside on Facebook.

Zuckerberg stuck to his talking points. Facebook's job was to balance free expression with safety. He didn't believe false information should be taken down. "Everyone gets things wrong, and if we were taking down people's accounts when they got a few things wrong, then that would be a hard world for giving people a voice and saying that you care about that," he said. His solution was to have users flag the false news, enabling the platform to make that content harder to find. But, he added, a lot of what could be described as false news was debatable.

"Okay, 'Sandy Hook didn't happen' is not a debate," Swisher said firmly. "It is false. You can't just take that down?"

Zuckerberg had prepared for the pushback. He said he also believed Sandy Hook deniers were wrong and that posts supporting that view were false. "Let's take this closer to home. I'm Jewish, and there's a set of people who deny that the Holocaust happened," he said, adding that he found this deeply offensive. "But, at the end of the day, I don't believe that our platform should take that down because I think there are things that different people get wrong. I don't think that they're *intentionally* getting it wrong." Facebook, said Zuckerberg, had a "responsibility to keep on moving forward on giving people tools to share their experience and connect and come together in new ways. Ultimately, that's the unique thing that Facebook was put on this Earth to do."

Swisher expressed her disagreement and went on to grill him repeatedly on how he knew that Holocaust deniers were not intentionally trying to mislead people about real-world events that had in fact transpired. Zuckerberg refused to back down. "He thought he was being so clever, making this point. He couldn't see he was being an intellectual lightweight and making an empty argument," Swisher later recalled. "I knew he was going to get skewered."

Within hours of the podcast's release, his remarks went viral.

"Mark Zuckerberg Defends Holocaust-Deniers," read the headline of one left-wing blog. National Public Radio summarized the story with the headline "Zuckerberg Defends Rights of Holocaust Deniers." Jewish groups in the United States, Europe, and Israel issued blistering rebukes, reminding Zuckerberg that anti-Semitism was an ongoing threat to Jews worldwide. Holocaust denial is a "willful, deliberate and longstanding deception tactic by anti-Semites," Jonathan Greenblatt, the chief executive of the Anti-Defamation League, said in a statement. "Facebook has a moral and ethical obligation not to allow its dissemination."

Hours later, Zuckerberg tried to clarify his comments in an email to Swisher, saying he did not mean to defend the intent of Holocaust deniers. Yet he hadn't stumbled into the topic; he had crafted the Holocaust denial response as an example of controversial speech that Facebook would defend in support of free expression. Since the 2016 election and the rise of fake news on Facebook, he had been grappling to come up with a coherent speech policy. Whetstone, who was jockeying to become Zuckerberg's top communications aide, encouraged him to be firm and draw a clear line on free expression. It was already a core tenet of his belief system, honed over the years through discussions with libertarians like Thiel and Andreessen. The patrons of Silicon Valley enjoyed defending absolutist positions, which they saw as intellectually rigorous. The holes in their arguments—the gray areas that people like Alex Jones or Holocaust deniers inhabited—were ignored. When Whetstone suggested that Zuckerberg bring up the most egregious example of speech he would allow, even if he personally disagreed with it, she was following the same line of argument that Thiel and others had used for years. It was a way for Zuckerberg to demonstrate his commitment to the idea that Facebook was a marketplace of ideas and that even uncomfortable speech had a place on the site.

Whetstone had joined the company less than a year earlier and had quickly assumed a senior role within the public relations team. The PR staff, numbering more than two hundred at that point, still struggled with crisis communications. Whetstone was no stranger to conflict. A former chief strategist to the British prime minister David Cameron, she had gone on to head up communications for Eric Schmidt at Google and then Travis Kalanick at Uber. Reporters feared her calls, in which she deployed every PR tactic in the book to dissuade them from running negative stories.

As soon as she arrived at Facebook, Whetstone implored Zuckerberg and Sandberg to change the PR culture. If Zuckerberg was serious about becoming a wartime leader, Whetstone told him, he needed a communications strategy to match. She insisted he go on the offensive.

Zuckerberg loved the idea. Personally, he found Holocaust deniers repulsive, which made them the perfect example for his purposes. By allowing them to create a community on Facebook, he was showing he could put his personal feelings and opinions aside and adhere to a consistent rule based on logic. He was confident that people would see his thinking as a difficult but necessary way to maintain the integrity of speech policy on Facebook. Several members of his PR staff pleaded with him to rethink the strategy. There was no need to invoke such an extreme case of what Facebook considered free speech; it would only blow up in his face. But he ignored their advice.

Zuckerberg viewed speech as he did code, math, and language. In his mind, the fact patterns were reliable and efficient. He wanted clear rules that one day could be followed and maintained by artificial intelligence systems that could work across countries and languages. Removing human beings, with their fallible opinions, from the center of these decisions was key: people made mistakes, and Zuckerberg, especially, did not want to be held responsible for making decisions on a case-by-case basis. In April, he had floated the idea of an outside panel that would decide on the toughest expression cases, a sort of "supreme court" that was independent of the company and had final say over appeals to Facebook's decisions.

In many areas of Facebook's content rules, Zuckerberg's worldview largely worked. The company banned violence, pornography, and terrorism—in those cases, AI systems caught and removed more than 90 percent of the content on their own. But when it came to hate speech, Facebook's systems proved reliably unreliable. Hate speech wasn't easily defined; it was constantly changing and culturally specific. New terms, ideas, and slogans were emerging daily, and only human beings deeply enmeshed in the world of the extreme, far-right movements cultivating such speech could keep up with its nuances.

"In Mark's ideal world there was a neutral algorithm that was universally applied and decided what was and wasn't allowed on Facebook," recalled one longtime Facebook executive who spent time arguing the merits of free speech with Zuckerberg. "He took it for granted that (a) that was possible, and (b) the public would allow and accept it."

The disastrous podcast interview clearly revealed that he had underestimated the complexities of speech, but he only doubled down on his position. Facebook could not be an arbiter of speech.

"He couldn't understand that speech isn't a black-and-white issue," said the executive who debated the subject with Zuckerberg. "He wasn't interested in the nuance, or the fact that when it comes to speech, there are certain things people simply feel, or know, are wrong."

To show the company's seriousness in fighting election interference, a war room had been set up on the MPK campus that served as a central command for a cross section of employees from the security, engineering, policy, and comms teams. Facebook's PR team invited select reporters to take a peek at the room to start planting stories ahead of the election. Some staff were embarrassed by the campaign, which they felt was too optimistic and designed for optics.

In case anyone was confused about the room's purpose, a paper sign was taped to the door with "War Room" typed in a large font. Inside, an American flag hung on one wall; another displayed digital clocks for Pacific Time, Eastern Time, Greenwich Mean Time, and the hour in Brasília, the capital of Brazil—the latter was of special interest to the group because Facebook was testing its new tools in the country, where elections were under way. Several oversize screens on the back wall displayed information being fed into a centralized dashboard Facebook had built to track the spread of content across the globe. On the fourth wall, television monitors were tuned to CNN, MSNBC, Fox News, and other cable networks.

In interviews with the press, Zuckerberg had described the fight against election interference as an "arms race" against foreign and domestic bad actors. The company had hired ten thousand new employees to help with security and content moderation, and its security team had dedicated more resources toward identifying foreign influence campaigns. Over the summer, the company announced that it had removed a network of Russian accounts aimed at influencing Americans ahead of the midterms, as well as a campaign of hundreds of Iranian and Russian accounts and pages engaged in disinformation efforts on behalf of their governments across parts of Europe and the Middle East.

"To see the way that the company has mobilized to make this happen has made me feel very good about what we're doing here," Samidh Chakrabarti, head of Facebook's elections and civic engagement team, said to reporters. "We see this as probably the biggest company-wide reorientation since our shift from desktops to mobile phones."

Zuckerberg and Sandberg had promised lawmakers that election security was their top priority. Among the ten thousand new employees were some prominent hires with big names in the intelligence and security worlds. They had approached Yaël Eisenstat, a forty-four-year-old veteran CIA officer, to lead the company's team to combat election interference on the company's business integrity side. Eisenstat was surprised by Facebook's overture; she didn't necessarily see herself as a fit for a private tech company with a dismal track record on election security. She had spent almost twenty years working around the globe for the U.S. government, fighting global terrorism. She had served as a diplomat and as a national security adviser to Vice President Biden and had worked in corporate responsibility for ExxonMobil.

Facebook had made Eisenstat the offer the same day in April that Zuckerberg testified before Congress on Cambridge Analytica. As she watched the hearing, she felt her interest being piqued by his promise that security in the midterms was his top priority, that he would throw Facebook's full resources behind efforts to ensure the integrity of elections. The lifelong government servant felt duty-bound to bring her skills to the social network to help protect American democracy, and she was promised a leading position in that fight. On her employment contract, her title was "Head of Global Elections Integrity Ops." She could hire as many people as she needed. She didn't have a set budget.

In a celebratory message to friends and colleagues, Eisenstat wrote, "Everything about this role speaks to the core of who I am. I spent much of my career and my personal life defending democracy, and now I get to continue that fight on the global stage by helping protect the integrity of elections worldwide. I cannot imagine a more perfect role to combine my public service mindset with the impact and scale of the world's largest social media platform," she said.

From the first day Eisenstat entered MPK, on June 10, however, the mismatch was clear. The motivational posters around the buildings and the displays of expensive art were off-putting. She felt she had entered a fog of propaganda. The perks—the free food, transportation, and gadgets—struck her as over the top.

On her second day, she was told that there had been a mistake in her title. She was not "head" of any team, but was instead given the nebulous title of "manager." Facebook executives had not only failed to tell the company about her new role but they had also taken it away without holding any conversation with Eisenstat. "I was confused. They were the ones that offered me this role. They were the ones who told me how much they wanted and needed me, and how important it was going to be to have someone with my perspective there," she recalled. "None of it made sense."

When she asked for details about the team and the efforts she would manage, she received a lecture on Facebook's values. One of her managers told her that titles didn't matter and that people moved around teams fluidly. Facebook would send her where she was most needed, the manager implied.

Eisenstat quickly realized that there were two parallel efforts under way at Facebook. She had been hired to work within Sandberg's team, which included policy and ads. But the elections integrity team, peopled by engineers and senior staff, reported to Zuckerberg's side of the company. Zuckerberg's team included Chakrabarti, who was the face of elections integrity work with reporters. Chakrabarti's team briefed Zuckerberg and other top executives on how they were preparing for the upcoming midterm elections.

For months, Eisenstat heard about meetings on election integrity but was told not to attend. One of her managers sent her to visit countries holding upcoming elections, but no one asked for a briefing or a memo when she returned. Through a chance encounter with Nathaniel Gleicher, Facebook's head of cybersecurity policy, she learned about the extent of Facebook's election work. "It was a short encounter, but I was excited to meet him and discover there was a place for me to contribute to Facebook's election effort. He even made a point of saying, 'Oh my god, you and I should be working together more,'" said Eisenstat. "It made me feel hopeful, that there were still people at Facebook that were trying to do the things I am passionate about."

But a second meeting with Gleicher did not materialize. Shut out of private discussions on the elections team, Eisenstat returned to a series of posts she had shared in the Tribe board used by the business integrity department. In one post, she had asked whether Facebook was applying the same standards to political ads as it was to organic content, or regular posts, on the platform. If Facebook was working to make sure that misinformation did not spread through its platform on pages and groups, wouldn't it make sense to ensure that a politician couldn't purchase an ad that deliberately spread false ideas or misinformation? In conversations with colleagues, Eisenstat argued that because Facebook had developed a system by which political ads could be microtargeted to a specific audience, such ads had the potential to become spreaders of false or misleading information. For instance, a candidate looking to smear his opponent's track record on the environment could target advertisements to people who had expressed an interest in recycling.

Eisenstat's message had generated conversation and interest across the company. Dozens of employees left comments, discussing what tools Facebook had built and what could be applied toward advertisements. Eisenstat was approached by a senior engineer who said that he was interested in beginning work as soon as possible on a tool that would allow Facebook to fact-check political ads, screening them for misinformation and other content that potentially broke the platform's rules. "A lot of engineers were energized by the post. We had been waiting to see if there was something connected to the elections that we could contribute to," said one engineer who recalled Eisenstat's message. "We were happy to volunteer time to help with her idea."

Then, just as quickly as the conversation had been sparked, it suddenly went dead. Eisenstat was not told how it happened or who among Facebook's executives had killed the initiative, but the message was clear: engineers no longer returned her emails, and interest in her idea evaporated.

Eisenstat had stumbled into a minefield. Since Trump had first upended American politics with his 2016 presidential campaign, he had posed a never-ending stream of challenges to Facebook and other social media companies. He posted statements and photos that included blatant mistruths. He shared conspiracy theories that had been widely discredited and deemed dangerous by law enforcement. And he endorsed political figures with ties to hate groups.

Zuckerberg had decided to give Trump special dispensation as an elected official, but a question loomed over how to handle advertisements bought and paid for by his campaign. Trump was the single-largest spender on political ads on Facebook. Mirroring his campaign, the ads often carried falsehoods or encouraged ideas that Facebook would otherwise have banned. It increasingly appeared that Facebook was accepting millions of dollars from Trump to spread his dangerous ideologies through Facebook tools that allowed him to reach highly targeted audiences.

Eisenstat thought Facebook was making a huge mistake by not fact-checking political ads. But by the fall of 2018, she could see that her direct managers were consistently sidelining her from any work connected to the upcoming midterm elections.

She decided to focus her energy on a project that would make the U.S. elections safer and more secure. She thought it was a no-brainer and exactly the sort of initiative Facebook had hired her to execute. This time, instead of posting about it on a message board for Facebook employees, she would work with her own team to come up with a prototype, research paper, and data points for a product guaranteed to ensure that American elections were more democratic.

Eisenstat knew that there was one point on which U.S. election law was absolutely clear: it was illegal to mislead or disenfranchise people from voting on Election Day. When she was in government, she had known people who focused on blocking efforts by groups that attempted to mislead voters by spreading false information, such as posting incorrect voting times or dates, or telling people they could vote by phone or email when they could not.

Over the course of several weeks, Eisenstat huddled with three of her colleagues, including an engineer and a policy expert. They built out a system that would locate and review any political ad on Facebook that potentially sought to disenfranchise voters. The system used many tools Facebook already had at its disposal, but it refocused them to look for the types of keywords and phrases commonly used to mislead and suppress voters. When they were done with the prototype, Eisenstat excitedly emailed her managers. "In the email, I was careful to stress that we collaborated across teams and that this effort came from the ground up, from our employees. I saw it as a win-win, protecting both the company and the public," she said. Instead, managers dismissed the protype as unnecessary and poorly conceived. "Right off the bat, in the email chain, they pushed back on what we were trying to do."

One of her managers accused Eisenstat of going behind their backs and ordered her to shut down the project. They warned her not to mention the idea to anyone. "I was repeatedly told not to go to anyone above my managers. I suppose I respected that because in the world I come from, in government, there is a strict chain of command," she said. "I didn't think to go around my managers."

Eisenstat went to HR and asked to be moved within the company. For weeks, they stalled, and then abruptly fired her, saying they couldn't find another role for her. Facebook, she thought, was still prioritizing the needs of the company over those of the country.

Eisenstat ultimately spent only six months at MPK. "I didn't join the company because my goal was to work at Facebook," she said. "I took the role because I thought it was one of the most important things I could do to protect democracy. I was wrong."

For weeks, the Facebook comms team had been tracking a story that reporters at the *New York Times* were chasing. On November 14, 2018, the piece, which outlined how Facebook had delayed, denied, and deflected the truth about Russian election interference over a two-year period, finally ran. The story was the first to detail when Facebook first discovered that Russia had been trying to influence U.S. voters during the 2016 elections, and it revealed how Facebook had hired an opposition research firm, Definers Public Affairs, to pursue its critics, including George Soros.

Just hours after the story was published, Zuckerberg and Sandberg met in the Aquarium with members of their communications and policy teams. Both had been warned that the story would be published that day, yet they appeared to have been caught flatfooted. Sandberg lambasted the team members. By the fall of 2018, Facebook had hired more than two hundred people to handle its press relations. Could none of them have done anything to influence the coverage? She asked the team how the reporters had been given the access to write the story. And then, when told that Facebook had given the *Times* very little access, she asked if a better strategy would not have been to invite the reporters in for interviews with her and Zuckerberg.

After the meeting, Sandberg, who had a number of friends in common with Soros, told her staff that she wanted to issue the strongest-possible denial in response to the piece. She was particularly concerned about the material on the Definers smear campaign. The public affairs group was led by Republican political operatives and was a classic opposition research firm, a dime a dozen in Washington, tasked with digging up dirt on opponents. It had sent research to journalists that showed financial ties between Soros and Freedom from Facebook, a coalition of groups that had criticized the company. The pitch was that the coalition wasn't operating out of a public interest but, rather, was a tool of the financier, who was known for his partisan, liberal views. But by choosing Soros as its target, it appeared that Facebook was playing into the tropes notoriously employed by fringe conservatives who had a history of smearing Soros with anti-Semitic conspiracy theories. A partner at Definers insisted the firm had no intention of attacking Soros on the basis of his background. But the timing of the revelations of the research couldn't have been less opportune. Weeks earlier, an explosive device had been sent to Soros's home in Westchester.

The statement Sandberg crafted with her staff read, "I did not know we hired them or about the work they were doing, but I should have." But the next day it was brought to her attention that she had been included on a number of emails highlighting the research Definers had done on behalf of Facebook. Several of those emails had just been shared with the *Times* reporters, who were preparing to write a follow-up story.

Schrage was asked to take responsibility for the public relations disaster. After taking a night to sleep on it, he let Sandberg know he would fall on his own spear. He had already announced that he was leaving his post at the company and would assume a role advising Zuckerberg and Sandberg. He drafted a blog post and showed it to Sandberg. In it, he shouldered the blame for having hired Definers: "I knew and approved of the decision to hire Definers and similar firms. I should have known of the decision to expand their mandate."

Sandberg was one of the first people to like and comment on the post. Facebook turned her comment into an odd epilogue of sorts, appending her response to the blog. In it, she thanked Schrage for sharing his thoughts and allowed that paperwork from the public affairs group may have crossed her desk. The responsibility was ultimately on her and other leaders, she wrote.

The carefully crafted comments from Sandberg rubbed a lot of people in the company the wrong way. It felt like a defensive, half-hearted apology. Calls began to circulate for her to step down from her position.

At a Q&A that month, an employee asked Zuckerberg whether he had considered firing any of his chief executives for their failings. Zuckerberg faltered momentarily, then responded that he had not. He gave Sandberg, who was seated in the audience, a small smile as he added that he still had trust in his leadership team to steer the company through the crisis.

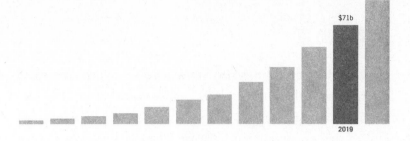

$71b

2019

Chapter 11

Coalition of the Willing

Zuckerberg was fuming. As he rode through the streets of Paris in a chauffeured black Mercedes V-Class MPV, he furiously scanned an article on his phone. The afternoon rain showers had let up, and the walkways along the Seine were teeming with pedestrians.

He was in France to meet Prime Minister Emmanuel Macron to discuss a surge in violence and hate speech on Facebook. It was the last leg of a global diplomatic offensive to defend the platform and to try to influence regulations under consideration in several nations. In the past five weeks, he had talked with government leaders of Ireland, Germany, and New Zealand.

Zuckerberg had aged visibly over the past year. His face was thinner, accentuated by his close-cropped haircut, and fine lines circled his red-rimmed eyes. His visit with Macron was the last hurdle to clear before he took a break from a grueling year of upheaval inside and outside of the company. He and Priscilla would celebrate Mother's Day at the Louvre before heading to Greece for their seven-year wedding anniversary.

But an op-ed published in the *New York Times* had interrupted his plans. Chris Hughes, Zuckerberg's Harvard roommate and a co-founder of Facebook, had delivered a scathing five-thousand-word rebuke of the company they had created together in their dorm room fifteen years earlier. In the piece, Hughes talked about their idealism in starting Facebook. But what they created, he wrote, had evolved into something much darker. The social network had become a dangerous monopoly, with 80 percent of the world's social networking revenue and a bottomless appetite for personal data. "It is time to break up Facebook," he said.

The core problem was Zuckerberg, Hughes asserted. He made the big decisions and held a majority stake of the company's voting shares. Mark was Facebook, and Facebook was Mark. And as long as he remained in charge, the only solution for the company's many problems was for the government to intervene and break the company into pieces.

"I'm angry that his focus on growth led him to sacrifice security and civility for clicks. I'm disappointed in myself and the early Facebook team for not thinking more about how the News Feed algorithm could change our culture, influence elections and empower nationalist leaders," Hughes concluded in the op-ed. "The government must hold Mark accountable."

Zuckerberg's expression hardened as he scrolled through the piece. He kept his gaze fixed on his phone, not uttering a word. After several minutes of silence, he looked up solemnly, with unblinking eyes. He told his aides he felt stabbed in the back.

And then he kicked into commander mode. How, he demanded, had the op-ed slipped through Facebook's army of PR staffers responsible for sniffing out negative articles? Whom had Hughes consulted in researching his critique? And what did he hope to achieve?

Within the hour, Facebook's PR team went on the offensive. Hughes hadn't been employed at Facebook for a decade, they told reporters; he didn't know how the company worked anymore. They questioned his motives and criticized the *Times*'s decision to publish the piece without giving Zuckerberg a chance to defend the company and himself. "Chris wants to get into politics," one Facebook flack informed a *New York Times* journalist.

But Hughes wasn't acting alone. He had joined a growing number of early Facebook executives, including former president Sean Parker, who were speaking out, admonishing the social network they had helped build and that had brought them individual wealth. He was also bringing energy to a movement in Washington to dismantle the social media giant. Political leaders, academics, and consumer activists were calling for the government to break off WhatsApp and Instagram, the fastest-growing services owned by Facebook. Two months earlier, Democratic presidential candidate Elizabeth Warren had vowed to break up Facebook and other tech giants if elected. Bernie Sanders and Joe Biden followed suit, promising tough scrutiny of Facebook, Google, and Amazon in their campaigns. Even President Trump, who used Facebook more effectively than any candidate in his presidential campaign, warned that the internet companies had too much power.

The meeting with Macron was tough, as expected, yet productive from the perspective of the long game: building a relationship with the French leader that could boost Facebook's reputation across Europe. Still, Zuckerberg continued to steam over Hughes's betrayal. Two days later, when a reporter for France 2 television news asked for his reaction to the op-ed, he publicly addressed it for the first time. It was a gray afternoon; rain streamed down the windows of the television station where Zuckerberg sat for the interview. His eyebrows furrowed as he looked toward the floor. "When I read what he wrote, my main reaction was that what he's proposing that we do isn't going to do anything to help solve those issues," he said, his voice rising slightly. He refused to refer to Hughes by name.

He didn't address any of the arguments raised by Hughes, such as the abuse of consumer privacy, the threat of disinformation to democracy, and Zuckerberg's too-powerful control. Instead, he warned of any intervention to dismantle Facebook's power. "If what you care about is democracy and elections, then you want a company like us to be able to invest billions of dollars a year like we are in building really advanced tools to fight election interference." A breakup would only make things worse, he explained.

Two months earlier, Zuckerberg had announced a radical shift in direction for the company, one that divided employees into warring camps. In a blog post titled "A Privacy-Focused Vision for Social Networking," he revealed that Facebook would focus on creating safe spaces for private conversations. In the past, users were largely encouraged to post on their own pages and those of their friends. The content then appeared in their News Feed, the constantly refreshing ticker that essentially functioned as a kind of virtual town hall. Now Zuckerberg wanted people to move to the privacy and security of the equivalent of their living rooms. It was a continuation of Facebook policies that had increasingly encouraged people to join groups.

Zuckerberg explained in the post that the company would also encrypt and link its three messaging services, Facebook Messenger, WhatsApp, and Instagram Messaging. It was billed as a move that would both help Facebook users streamline their messages in one central place and offer extra protection by encrypting those messages. "I believe the future of communication will increasingly shift to private, encrypted services where people can be confident what they say to each other stays secure and their messages and content won't stick around forever," Zuckerberg wrote. "This is the future I hope we will help bring about."

The announcement caught some of his executives by surprise. Zuckerberg had consulted only a few leaders of the M-Team when formulating the plan. The "pivot to privacy," as it would come to be known, triggered alarm among Facebook's security experts and some executives, who felt the focus on encryption and groups carried potentially dangerous consequences. Zuckerberg was effectively weakening Facebook's ability to serve as a watchdog over its own technologies.

Rita Fabi, a director of online safety, was among the most vocal opponents of the plan. Since joining Facebook in 2010, she had helped oversee cases of harmful content, including the abuse of children and sex trafficking. Her small group within the security team worked to catch predators. It dealt with the most disturbing content on the site, and its members were regarded by many coworkers as among the most committed and diligent employees at the company. Fabi put together legal cases against predators that Facebook had found, and she worked closely with law enforcement, sharing evidence and updating them on ongoing investigations through the National Center for Missing and Exploited Children. She had helped provide evidence used to prosecute numerous predators. In the offices where Facebook's security team sat, a wall was dedicated to the mug shots of the criminals the company had helped put away. They called it their "scalp wall."

Fabi wrote a memo and posted it to Facebook's Workplace group for all employees, tagging Zuckerberg and other executives. Because of the sheer volume of content created every day on the site, Facebook relied in large part on users to report suspicious activity, but private groups of like-minded users would make that much harder. Similarly, Fabi pointed out, encrypting messaging would allow criminals to hide in plain sight of Facebook's security team. Predators could join a small, private Facebook group for fans of a teen pop idol and more easily lure members of the group over encrypted chat. A pedophile could form his own tight-knit Facebook group for other child molesters to share tips on how to prey on children and could use code words to evade detection by Facebook's security team.

Chris Cox was also alarmed by the gaping security holes created by the privacy initiative. Of all the M-Team members, Cox had a particularly close relationship with Zuckerberg; they often socialized in the evenings and on weekends. Beloved by employees, he served as the internal evangelist for the company's mission, welcoming new hires at orientation, but he also had credibility as a talented coder—he was one of the original fifteen engineers at the company back when it was The Facebook.

But in a rare clash, Cox disagreed with Zuckerberg's initiative. For months, he had been quietly protesting in private conversations with Zuckerberg the plan to encrypt and merge the messaging services. Many of his arguments echoed Fabi's concern about the potential for criminal activity. Private groups were also harder to police for the spread of disinformation, he pointed out: Facebook was setting the stage for conspiracy theorists, hate groups, and terrorists to organize and to spread their messages under a cloak of darkness, using the platform's tools.

Zuckerberg heard out Cox's position, but told his longtime friend and confidant that he felt the private groups were the way forward. He wouldn't back down. Fabi quit, and then Cox, a thirteen-year veteran of the company, handed in his resignation.

Cox's departure was a major blow. He was often described as the "soul" of the company and was seen by many as the heir apparent, if Zuckerberg ever decided to step away from Facebook. He must have been offered a CEO position at another company, or a position in government, employees speculated when they first heard the news.

"People couldn't imagine Chris leaving for anything short of some dream job somewhere else," said one engineer who had worked with Cox for years, across several departments. "When people found out he had left because he disagreed with the direction Mark was taking the company, it was, well, just crushing. It felt like our parents were getting divorced."

On his Facebook page, Cox shared the news in a post that included a picture of himself with his arm around Zuckerberg. The privacy announcement marked a new chapter in product direction, he wrote, and Facebook "will need leaders who are excited to see the new direction through."

Cox added that he had given his last orientation that week, and he repeated a line to the new employees that he had delivered hundreds of times: "Social media's history is not yet written, and its effects are not neutral."

Even after Chris Hughes left Facebook, he and Zuckerberg had remained friends. They attended each other's weddings and kept each other updated on milestones in their lives. In the summer of 2017, Hughes dropped by Zuckerberg's home in Palo Alto to visit with Mark, Priscilla, and their daughter Max, then a toddler. They caught up on family and Facebook and life on opposite coasts.

Hughes had been watching Facebook's evolution with unease. Russian election interference, the Cambridge Analytica scandal, and the rise of hate speech transformed his unease into anger and guilt. Then, on April 24, 2019, he read Facebook's announcement in an earnings report that the FTC was poised to fine the company as much as $5 billion for the Cambridge Analytica data privacy breach. It would be a record penalty, exponentially bigger than any past action against a Silicon Valley company, and a rebuke of the practices at the heart of Facebook's juggernaut behavioral advertising business. The biggest data privacy fine against a tech company until then was the penalty imposed on Google in 2012, for $22 million. Hughes considered the looming FTC settlement a just punishment. "Oh my god," he muttered to himself, reading the announcement in his office loft in Manhattan's West Village. "This is it."

Curious to see how far Facebook's stock would plummet upon news of the expected FTC fine, Hughes opened a Yahoo finance tab. To his shock, the price was soaring. Contrary to what he had predicted, the fine was welcome news on Wall Street; investors were thrilled that the FTC inquiry had been resolved. And Facebook had announced its quarterly earnings report that same day: ad revenues were up 26 percent; the company had cash reserves of $45 billion and annual revenues of $71 billion. The record fine, just 3.5 percent of revenues, would hardly be a setback.

"I was outraged," Hughes said. "Privacy violations were just the cost of doing business."

In 2016, Hughes had founded a progressive think tank and advocacy group, Economic Security Project, which proposed federal and local tax reforms to give guaranteed monthly income to the poorest Americans. He was part of a growing movement among left-leaning academics and politicians against the concentration of power held by Big Tech; the movement pointed to Facebook, Amazon, and Google as the winner-take-all titans of a new gilded age. The chorus was growing: advocates Elizabeth Warren and Bernie Sanders were joined in their protest of Big Tech's monopolization by the progressive former labor secretary Robert Reich, George Soros, and others.

But Hughes acknowledged that the seeds of Facebook's problems were there from the beginning: the urgent-growth mindset, the advertising business model, and Zuckerberg's centralized power. No one individual, not even the friend he described as "a good person," should have that much control over such a powerful institution. Facebook's repeated data privacy violations were symptomatic of a bigger problem. Even a record penalty by the FTC wouldn't deter the company's dangerous business practices. With

2.3 billion users and a market valuation of $540 billion, it had become too big to be restrained by regulations alone.

"Obviously, my whole life has been changed by one of the biggest monopolies in the United States today," Hughes said. "And so, I needed to take a step back and wrestle with the fact that Facebook has become a monopoly, and name that and explain why that happened, how that plays into all of its mistakes, and the growing cultural and political outrage," he recalled. "And then also understand what can be done about it."

Two months before Hughes's op-ed, Zuckerberg had announced the plan to break down the walls between messaging services in Facebook, Instagram, and WhatsApp. The company would stitch together more of the back-end technologies of the apps and allow users to message and post content across them. Making the apps "interoperable" would be a boon for consumers, he said. If a user found a small business on Instagram, they would be able to use WhatsApp to message that business. Engineers would encrypt messaging on all the apps, providing more security to users.

"People should be able to use any of our apps to reach their friends, and they should be able to communicate across networks easily and securely," Zuckerberg wrote. But in reality, he was fortifying the company. Blue, Instagram, and WhatsApp were three of the top seven apps in the United States, with a combined 2.6 billion users globally. Facebook could merge data from each app and glean richer insights into its users. Engineering teams would work more closely to drive greater traffic between the apps, keeping users within the Facebook family of apps longer and more exclusively. It was like scrambling the apps into an omelet.

Zuckerberg was backsliding on concessions he had made to government officials to get the original buyout deals through regulatory review. He was also reversing promises made to the founders of Instagram and WhatsApp to keep the apps independent. In 2012,

he assured Kevin Systrom and Mike Krieger that he wouldn't interfere with the look, feel, or business operations of Instagram. The
founders were obsessed with the style and the design of Instagram
and feared that Facebook, which had barely evolved its look from
its earliest days, would mess with Instagram's DNA. After the
merger, Instagram had stayed at arm's length from Zuckerberg and
Sandberg, in its own space at MPK, one that was decorated differently from the rest of Facebook.

Facebook's vice president of development, Amin Zoufonoun,
said the promises made to Instagram's founders were restricted to
the brand—that Facebook wouldn't interfere with the look or feel
of the photo-sharing app—and from the beginning of merger discussions, there was talk about how Instagram "would benefit from
Facebook's infrastructure."

"Over time, Instagram increasingly shared more Facebook infrastructure as it became clear that we needed a more unified strategy
across the company," Zoufonoun said. But other employees described a more fraught relationship.

Instagram was popular among a younger demographic, while
Blue's average was skewing older. The contrast in demographics
irked Zuckerberg, who feared Instagram could cannibalize the Facebook app. Within six years, he began to interfere with technological
decisions at Instagram. He tried to introduce a tool that would drive
Instagram traffic to Facebook. Sandberg, for her part, insisted on
more ads on the app. Instagram introduced its first ads in November 2013 for select big brands. In June 2015, the photo-sharing app
opened to all advertisers. Instagram had become hugely successful
in the seven years under Facebook's control, with an estimated value
of $100 billion and a billion users. "Mark saw Instagram's poten-
tial from the start and did everything he could to help it thrive,"
Adam Mosseri, the head of Instagram, maintained. "It's not logical
for Facebook to do anything that would dampen this success."

But for Instagram's founders, Zuckerberg and Sandberg's involvement became overbearing. Systrom and Krieger quit in September 2018, saying they were "planning on taking some time off to explore our curiosity and creativity again."

The story was similar with WhatsApp. Jan Koum and his co-founder, Brian Acton, had sold WhatsApp on the condition that the company would operate independently from the rest of Facebook. Zuckerberg had assured Koum, who feared government surveillance via apps, that he would make privacy a priority. Both conditions were selling points to regulators as well. Acton claimed that Facebook trained him to explain to European regulators that merging would be challenging. "I was coached to explain that it would be really difficult to merge or blend data between the two systems," he said. But eventually, Zuckerberg and Sandberg imposed themselves on WhatsApp and pressured Acton and Koum to monetize the messaging app. Acton said that, in a meeting with Sandberg, he proposed a metered pricing model that charged users a small fraction of a penny for a certain number of messages. Sandberg shot down the idea, claiming, "It won't scale."

In December 2017, Acton quit in protest over Zuckerberg and Sandberg's plans to introduce targeted advertising to users. He became a billionaire from the deal; later, he conceded that he'd been naïve and regretted trusting Zuckerberg. Business came first, and they were good businesspeople. "At the end of the day, I sold my company," Acton said in an interview with *Forbes*. "I sold my users' privacy to a larger benefit. I made a choice and a compromise. And I live with that every day."

Tim Wu and Scott Hemphill, law professors at Columbia and New York University, respectively, viewed the integration of the apps as a ploy to ward off antitrust action. Zuckerberg could argue that the process of breaking them up would be too complicated and harmful to the company—you couldn't actually unscramble the eggs in an omelet. Facebook would have good reason to pursue such a strategy: regulators were reluctant to review past mergers, especially deals that hadn't been blocked. Unwinding mergers was a messy task that could cause more harm than good.

But the law professors had been studying Facebook's history of acquisitions. They saw Zuckerberg's announcement as a defensive measure to protect his empire. Wu had an insider's understanding of government. He had worked at the FTC, in the Obama White House, and for the New York attorney general. He was best known for coining the term *net neutrality*, the principle that telecommunications providers should keep their high-speed internet networks open for all content. Google and Facebook cheered his anti-telecom theories and joined his push for regulations on broadband service providers. In recent years, however, he had flipped sides and begun to rail against the power of those tech partners.

Facebook's acquisitions of Instagram and WhatsApp were part of a pattern to create a social media monopoly and then to maintain that monopoly by buying or killing off competition, the professors argued. They saw a historical parallel in Standard Oil, which became a monopoly during the industrial age through acquisitions, eventually overseeing more than forty companies. The oil trust, established in 1882 by John D. Rockefeller, then maintained its monopoly by colluding with partners and squashing competition, ultimately putting itself in a position to own or influence every part of the supply chain. At its peak, the company controlled almost all oil production, processing, marketing, and transportation in the United States.

By 2019, Facebook had acquired nearly seventy companies, the vast majority of the deals valued at less than $100 million and not subject to regulatory review. When the Instagram deal was reviewed by the FTC in 2012, Facebook argued that the photo-sharing app wasn't a direct competitor. The company made a similar argument in defense of the WhatsApp acquisition in 2014: the app didn't compete with Facebook's core business as a social network. The regulators didn't block the deals because there wasn't enough evidence that Facebook was eliminating competitors. Instagram and WhatsApp didn't have advertising businesses at the time; the FTC didn't consider how, in the process of absorbing the apps, Facebook would feed more data into its ad business and cement its dominance in social media. "The lessons from history were all there, and yet we were going down the same path of the Gilded Age," Wu said. "That era taught us that extreme economic concentration yields gross inequality and material suffering, feeding an appetite for nationalistic and extremist leadership."

Facebook also tried to squash competitors, the law professors argued. In 2013, internal emails revealed that Facebook cut off Twitter's short-video service, Vine, from a friend-finding feature of the Open Graph, according to a document released by the UK's Digital, Culture, Media and Sport Committee, which was investigating Facebook for alleged data privacy violations. Industry analysts maintained that the decision hurt Vine, which shuttered in 2016. "The files show evidence of Facebook taking aggressive positions against apps, with the consequence that denying them access to data led to the failure of that business," asserted Damian Collins, the committee member who released the documents.

In conversations with acquaintances and competitors, Zuckerberg used the line "Now you know who you are fighting," a quote from the movie *Troy*, according to one friend. Like the Greek hero Achilles, who speaks the line in the movie, he was signaling his intent to demolish his opponent.

Wu put many of the ideas he and Hemphill had developed in *The Curse of Bigness: Antitrust in the New Gilded Age*, which he published in November 2018. Hughes had read Wu's book during a family vacation in Mexico soon after it was published, and when he returned to New York, the two met and began to exchange ideas. Then they started reaching out to allies from their networks of Democrats and business leaders.

In late January 2019, Wu ran into Jim Steyer, Common Sense's founder, at an after-party at the Sundance Film Festival in Park City, Utah. Wu ran his antitrust theories about Facebook by Steyer, who found them galvanizing.

They decided to bring in other agitators, including Steyer's Exeter classmate Roger McNamee, an early Facebook investor who was writing a scathing critique of the company called *Zucked*. They also sought advice from Common Sense's policy adviser and former chief of staff to Vice President Joseph Biden, Bruce Reed, who had fought against Facebook for a privacy law in California.

Sentiment against the company was at an all-time high. Civil rights organizations and consumer privacy advocates decried the company as a gross offender of human rights and data privacy violations. Lawmakers on both sides of the aisle were regularly publishing critical statements; Trump had accused Facebook and Twitter of censorship. Wu's group was opening another front against Facebook, with an aim to establish a clear antitrust case against the company. "I call us the Coalition of the Willing," Wu said, a nod to the U.S.-led multinational forces in the George W. Bush administration's invasion of Iraq.

Wu and Hemphill created a slide deck with their arguments, and in February, they took their show on the road. They began a series of meetings with officials in the offices of eleven state attorneys general, including New York, California, and Nebraska. Hughes joined Wu and Hemphill on a conference call for a meeting with the Department of Justice and the FTC, where they implored the nation's top antitrust officials to break up Facebook.

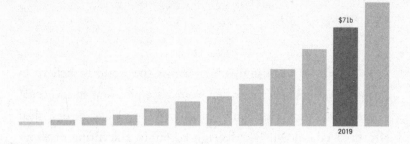

$71b

2019

Chapter 12

Existential Threat

The day before Hughes published his op-ed, Sandberg met with House Speaker Nancy Pelosi. The appointment capped two days of difficult meetings with Republican and Democratic lawmakers in charge of the Commerce and Intelligence Committees in both chambers. The legislators grilled Sandberg about privacy abuses and efforts to prevent disinformation during the 2020 elections and demanded signs of progress.

It was a trying period for Sandberg. Her work responsibilities were crushing: friends said she was feeling tremendous pressure, and some guilt, for the cascade of scandals confronting the company.

When Pelosi appeared in the sitting area for guests, Sandberg greeted her with a smile. The Speaker responded coolly, inviting Sandberg to sit with her on the couches in the office meeting area. The tense mood was in stark contrast to Sandberg's visit to Pelosi's office in July 2015, to discuss women in leadership. Then, just months after Goldberg's death, Sandberg had begun to travel again and was embraced by lawmakers. "What a pleasure to visit with Sheryl Sandberg and her team in my Capitol office," Pelosi had effused in a Facebook post accompanying a photo of the two women in front of a large gilded mirror and the American flag. "Thank you, Sheryl, for inspiring women across the world to believe in themselves. We know that when women lean in, women succeed!"

Now, four years later, Sandberg ignored the chill as she described efforts to take down fake foreign accounts, the hiring of thousands of content moderators, and the implementation of AI and other technologies to quickly track and take down disinformation. She assured Pelosi that Facebook would not fight regulations. She pointed to Zuckerberg's op-ed in the *Washington Post* in April that called for privacy rules, laws requiring financial disclosures in online election ads, and rules that enabled Facebook users to take their data off the social network and use it on rival sites.

The two talked for nearly an hour. Sandberg was admitting Facebook had problems, and the company appeared to be at least trying to fix them. Pelosi was still on guard, but the efforts appeared to be a step forward. Finally. They seem to be getting it, Pelosi told her aides after the meeting.

Two weeks later, a video featuring the Speaker went viral.

On Wednesday, May 22, 2019, a video clip appeared on Facebook showing the Speaker sitting onstage at a conference held a day earlier by the Center for American Progress, a liberal think tank based in Washington, DC. In the video, Pelosi is clearly in a good mood. Earlier in the day, she had tussled with Trump at the White House over her push for impeachment proceedings, and she was recounting the event for a packed audience at the downtown Renaissance hotel. The crowd rose to their feet to applaud her.

But in the video posted of the event, something appeared to go very wrong. The Speaker's words began to slur: "We want to give this president the *opp-or-tun-i-ty* to do something historic," she said, painfully drawing out each syllable. She appeared inebriated. As she recounted how the president had stormed out of the meeting, her voice went into vocal slow motion. "It was very . . . very . . . very . . . strange."

The video was quickly reposted and shared. On a page called "Politics Watchdog," it attracted two million views and was shared tens of thousands of times. It racked up more than two thousand comments, with users calling Pelosi "drunk" and "deranged." From there, it was shared to hundreds of private Facebook groups, many of them highly partisan pages. Within twenty-four hours, Trump's personal lawyer and the former mayor of New York City, Rudy Giuliani, had tweeted the link, along with the message "What's wrong with Nancy Pelosi? Her speech pattern is bizarre." The hashtag #DrunkNancy began to trend.

But Pelosi was not intoxicated. She doesn't drink alcohol. In the original broadcast of the event on the public cable channel C-SPAN, Pelosi was lucid and spoke at an even pace. The video circulating on Facebook and other social media had been manipulated: the original footage had been doctored through simple audio editing techniques that slowed down Pelosi's speech to around 75 percent of normal speed. It was a frighteningly cheap and easy tweak that opened a new front in the tech industry's battle against political misinformation.

The private Facebook groups Zuckerberg had championed as part of his pivot to privacy two months earlier were the ones now spreading the viral video. Within the confines of the small groups, Facebook users not only joked with one another about how to edit the video but also shared tips on how to ensure that it would go viral and reach the maximum number of people.

Pelosi's staff were livid. They questioned why the social networks had allowed the altered video to spread. The Speaker was also frustrated. The internet companies had assured her that they'd been working to clean up their sites. Not even three weeks earlier, Sandberg had sat in her office and said as much.

Call the companies and tell them to take down the video, she told her aides.

Within hours, YouTube had removed the clip, citing it as a violation of its policy against misinformation. In the meantime, the media had picked up on the doctored video. Stories in the *Washington Post* and other outlets broke down the technical manipulation involved. Academics and lawmakers declared it a litmus test for social media platforms ahead of the 2020 presidential election.

In response to the update on YouTube's removal of the video, Pelosi asked her staff, "What about Facebook?" The video was still up on several websites, including extreme right-wing news sites and fringe message boards, but Facebook was where it was getting significant traction.

Facebook, her aides said, had gone silent. No one was picking up their calls.

Pelosi was dumbfounded. Her office had particularly strong ties to the company. Catlin O'Neill, Pelosi's former chief of staff, was one of Facebook's most senior Democratic lobbyists. Other staff came from the Obama administration and had worked for Senator Feinstein and former senator John Kerry. It was beginning to seem like perhaps those ties were meaningless.

Democrats had quickly jumped to defend Pelosi. Rep. David Cicilline of Rhode Island tweeted soon after the video was posted, "Hey @facebook, you are screwing up. Again. Fix this now!" Sen. Brian Schatz called out the platform for hypocrisy: "Facebook is very responsive to my office when I want to talk about federal legislation and suddenly get marbles in their mouths when we ask them about dealing with a fake video. It's not that they cannot solve this; it's that they refuse to do what is necessary."

On Thursday, Zuckerberg and Sandberg in Menlo Park and Kaplan in Washington, DC, met over videoconference to come up with a response. The fact-checkers and the AI that Sandberg had touted to Pelosi hadn't flagged the video for false content or prevented its spread. It was easy to foil Facebook's filters and detection tools with simple workarounds, it turned out. Just a few months earlier, the platform's systems had failed to find and remove horrifying first-person footage of a gunman in Christchurch, New Zealand, who, on March 17, opened fire on worshipers gathered in two mosques, killing fifty-one people. It was a video shot and designed to go viral, and the gunman had included in it several references to conspiracy and hate groups on Facebook whose members he knew were sure to promote the footage of his carnage once he posted it on the site.

For twenty-four hours, Facebook battled to keep up with the spread of copies of the video that had been posted on the site. Of the 1.5 million copies people tried to upload to the site, Facebook caught and removed 1.2 million. That left 300,000 versions of the video, some of which stayed up on the platform for minutes, others for hours. While the company boasted that it had been able to remove over 90 percent of the video uploads, the remaining copies exalting a brutal massacre were far too many. And it exposed how users were able to get around Facebook's systems for catching and removing videos. By including just small alterations, like slowing down the video by a millisecond or putting a watermark over part of it, people were able to fool Facebook's vaunted AI. Some users simply took new videos of the video. The videos surfaced at the top of News Feeds and generated millions of comments and shares within private groups.

The massacre at Christchurch demonstrated how Zuckerberg's reliance on technology solutions was insufficient. Government officials in France, the United Kingdom, and Australia decried Facebook for allowing its tools to be weaponized. Several nations proposed laws on speech after the massacre. Macron and Jacinda Ardern, the prime minister of New Zealand, took a lead and called for a global summit to put a stop to violent and hateful content online.

But the doctored video of Pelosi revealed more than the failings of Facebook's technology to stop its spread. It exposed the internal confusion and disagreement over the issue of controversial political content. Executives, lobbyists, and communications staff spent the following day in a slow-motion debate. Sandberg said she thought there was a good argument to take the video down under rules against disinformation, but she left it at that. Kaplan and Republican lobbyists encouraged Zuckerberg's "maximum openness" position on speech and stressed that it was important to appear neutral to politics and to be consistent with the company's promise of free speech.

The discussions became tortured exercises in "what-if" arguments. Zuckerberg and other members of the policy team pondered if the video could be defined as parody. If so, it could be an important contribution to political debate; satire had always held a critical place in political discourse. Some communications staff noted that the same kind of spoof of Pelosi could have appeared on the television comedy show *Saturday Night Live*. Others on the security team pushed back and said viewers clearly knew that *SNL* was a parody show and that the video of Pelosi was not watermarked as a parody. "Everyone was frustrated by how long it had taken," one employee involved in the discussions explained. "But when you are trying to write a policy for billions of people, you have to abstract that and think of all unintended consequences."

In the meantime, Monika Bickert and the content policy team began a technical analysis of the video. They wanted to determine if it met the company's threshold definition for a "deep fake" video, a clip that used machine learning or artificial intelligence to alter content synthetically. The cheaply edited video didn't meet the threshold.

On Friday, forty-eight hours after the video first surfaced, Zuckerberg made the final call. He said to keep it up.

Pelosi wasn't nearly as critical of Facebook as some of her colleagues had become; she had always viewed tech, in the backyard of her California district, as an important industry for the economy. But the damage had been done. "Do not take calls, do not take meetings, do not communicate at all with Facebook," Pelosi told her staff. The company had lost support from the most important holdout within the Democratic Party.

Facebook's lobbyists insisted that Zuckerberg and Sandberg had to call Pelosi personally to explain the company's decision. She refused their calls. The following week, she delivered a blistering assessment of the social network. "We have said all along, 'Poor Facebook, they were unwittingly exploited by the Russians.' I think wittingly, because right now they are putting up something that they know is false," she declared in a San Francisco public radio interview. Facebook, she added, was contributing to the problem of disinformation on its site. "I think they have proven—by not taking down something they know is false—that they were willing enablers of the Russian interference in our election."

Privately, Sandberg told people close to her that she was torn up by Zuckerberg's decision. She felt there was a strong argument that the doctored video violated the company's rules against misinformation, and the refusal to take it down cancelled out her efforts to win back trust in the company. And yet she didn't push back in the videoconferences, some colleagues noted. It frustrated those who felt she was in the best position to challenge Zuckerberg. Within the company, she clearly followed the command of her CEO. "Only one opinion matters," she often told aides.

Weeks later, during a panel at the Cannes Lions marketing conference in Cannes, France, Sandberg was asked about the video. She allowed that the decision to keep it up was "hard and continues to be hard." But she did not mention Pelosi by name, and she echoed her boss's now-familiar position on speech. "When something is misinformation, meaning it's false, we don't take it down," she said, wearing a flat, matter-of-fact expression. "Because we think free expression demands that the only way to fight bad information is with good information."

Zuckerberg worried that the breakup proposition was gaining momentum. In the late spring of 2019, a Facebook researcher ran a series of public surveys on opinions of regulating and breaking up the major tech companies. Respondents said that of all the options, they were most comfortable with a government mandate to unwind Facebook.

The company deployed Schrage's replacement, Nick Clegg, to bat down the idea. Clegg wrote his own op-ed in the *New York Times* to refute Hughes's characterization of Facebook as a monopoly, pointing to competition from Snapchat, Twitter, and the rising threat of the Chinese short-video service TikTok. He warned, "Chopping a great American success story into bits is not something that's going to make those problems go away."

Unlike Schrage, Sir Nicholas Clegg was a natural public speaker and an ambassador for the company with global leaders. The former deputy prime minister had exited British politics in defeat. As the leader of the centrist Liberal Democrats, Clegg had joined Prime Minister David Cameron's Conservative Party in a coalition focused on austerity, but he attracted criticism for his handling of university tuition fees, which were introduced under his watch, and for his emphasis on reducing debt through cuts to public spending. But he was a fresh face to U.S. politicians, and a gifted orator. He advised Zuckerberg and Sandberg to take a more proactive stance politically. He said regulations were inevitable, so Facebook needed to lead tech companies in a public campaign to create light-touch rules, instead of more stringent laws.

Clegg and Kaplan hit it off. They had a similar view on free expression, and both viewed Facebook's business through the lens of politics. They came up with a strategy, the result of which was to deflect scrutiny by pointing to the economic and security threats posed by China. Zuckerberg had all but given up entering the Chinese market; in an internal post in March, he said he would not put data centers in countries with censorship and surveillance practices. Zuckerberg and Sandberg began to criticize China in speeches and interviews.

Kaplan took their message to Washington. Over the spring and summer, as he made the rounds of tech lobbying groups, he and his lobbyists proposed the creation of a separate trade group in support of domestic technology companies, to be led by former U.S. military officials. The underlying idea was that the success of American technology companies was the best defense against China's military and industrial might. Facebook's lobbyists approached leaders of the Internet Association, a trade group, to see if any members—such as Google, Amazon, and Microsoft—would join the new trade group he called "American Edge." He couldn't get support. "No one wanted to associate with Facebook," one industry executive said. Kaplan created American Edge anyway, with Facebook as its sole corporate member.

In the span of two weeks in early June, three investigations into the platform's practices were launched. The Federal Trade Commission opened an investigation into Facebook's monopoly power. Led by New York, eight state attorneys general also began their own joint antitrust investigation of the social network. The House Judiciary Committee's antitrust subcommittee began a separate inquiry into Facebook and other tech giants.

The multiple government inquiries, extraordinary for any corporation, caught many in the Washington Facebook office unaware. But Zuckerberg wasn't slowing down. In June, he announced Libra, a blockchain currency system that could replace regulated financial systems. The virtual currency plan drew immediate criticism from global regulators, who warned that a system run by a private company—especially Facebook—could harbor illegal activity. Congress promptly called for the company to appear in hearings. Kaplan responded by throwing more resources into the Washington office. Between July and October, he hired five additional lobbying firms to help fight against the government's intervention on Libra.

The sense that the company was coming under threat was filtering down the ranks. Employees at MPK were growing concerned about the antitrust action. In two Q&A meetings in July, they asked Zuckerberg about the drumbeat in Washington to break up Facebook, including the specific promise by presidential candidate Elizabeth Warren to dismantle the company.

If Warren were elected, the company would brace for a fight for its life, Zuckerberg said. "At the end of the day, if someone's going to try to threaten something that existential, you go to the mat and you fight."

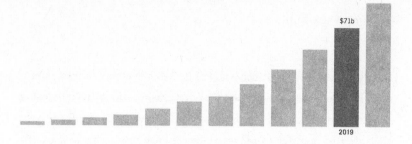

$71b

2019

Chapter 13

The Oval Interference

Zuckerberg slipped into the Oval Office on the afternoon of September 19, 2019. The appointment wasn't on the official White House schedule; the president's press pool only reported that they "did not lay eyes on the president" all day. The meeting of two of the world's most powerful men was held in secret.

Trump leaned forward, resting his elbows on the ornately carved nineteenth-century *Resolute* desk. As he boasted about the performance of the economy under his administration, a jumbo-size glass of Diet Coke collected condensation on a coaster in front of him. Zuckerberg sat on the other side of the desk, in a straight-back wooden chair wedged between Kaplan and Jared Kushner, Trump's son-in-law and senior adviser. Dan Scavino, Trump's director of social media, sat at the end of the row. Behind Zuckerberg, a small gold statue of Poseidon, the Greek god of the sea, rested on the marble fireplace mantel.

The president did most of the talking. For all his team's success using Facebook, Trump was skeptical of the social network. He had claimed that it juiced its algorithms to censor his supporters. He also held a grudge about Silicon Valley's overt support of Hillary Clinton during the 2016 election, including the endorsements from

Sandberg and other Facebook executives. "Facebook was always anti-Trump," the president had posted on Facebook in September 2017. Trump and his campaign posted frequently on Facebook, though the president preferred Twitter for raging against political rivals and the media. But he needed the platform to reach the public as much as Facebook needed to be at the center of public conversation, which for the past three years had been consumed by the president.

Zuckerberg's introduction to Trump's White House had come through Kaplan and Peter Thiel. Zuckerberg had first gotten to know Kushner, who graduated from Harvard the year Zuckerberg began. He scheduled an appointment with the senior adviser right before the Oval Office meeting to deliver a compliment: "You were very good on Facebook," he told Kushner, who had helped lead Trump's election campaign.

The goal of the meeting was to neutralize Trump's animosity toward Facebook, so Zuckerberg offered a vanity gift of sorts. His team had run the numbers using proprietary internal data, and the president had the highest engagement of any global leader on Facebook, Zuckerberg told him; Trump's personal account, with 28 million followers, was a blowout success. The former reality show star visibly warmed up.

Zuckerberg had prepped on a wide array of topics Trump was likely to raise, including his animus toward China. It was the president's hobbyhorse. Trump was engaged in a trade war with the country, part of a nationalist agenda to shore up the power of American companies against the rising dominance of Chinese global business giants. He had banned the Chinese telecommunications companies Huawei and ZTE from selling equipment in the United States and pressured allies in Europe to reject Chinese tech and telecom equipment and services. When the opportunity arose, Zuckerberg pounced, chiming in that Facebook's rivals in

China were dangerously ascendant. Apps like TikTok and WeChat were among the top in the world, racking up more downloads in the iTunes and Android stores than most of the American competition. The spread of China's government-sponsored tech sector threatened America's leadership in innovation and technology, the two men agreed.

The hour-long discussion ended on a genial note. Later in the day, Trump disclosed the meeting on Facebook and Twitter, posting a photo of the two men shaking hands, a wide smile on the CEO's face. "Nice meeting with Mark Zuckerberg of @Facebook in the Oval Office today," read the caption.

The Oval Office encounter was considered a coup by Kaplan and Nick Clegg, who felt strongly that Facebook's leaders needed to interact more with the current administration. Central to the strategy was a meeting with Trump. Zuckerberg had been wary of engaging directly with the president, even as his counterparts at other businesses had made the pilgrimage to the White House. Privately, he told close aides he was disgusted by Trump's detention of illegal immigrants and his anti-immigrant rhetoric. Zuckerberg had long supported immigrant rights, including Dreamers, the children of undocumented immigrants. Priscilla was the daughter of ethnic Chinese refugees from Vietnam; when she was teaching in East Palo Alto, she suggested Zuckerberg tutor students at her high school. He had been deeply affected by the experience and had become a champion of immigration. In 2013, Zuckerberg formed Fwd.us, a lobbying group for pro-immigration legislative reforms, including higher quotas for skilled workers. Aside from comments made right after the election, however, Zuckerberg didn't speak out about Trump's stance on immigration.

The cultural rift between the Washington office and MPK was widening. The Oval Office meeting upset many employees in Menlo Park, but to their counterparts in Washington, Kaplan was simply following the lobbying playbook used across corporate America. His main goal was to preserve the status quo and to prevent regulations that would obstruct the company's profit machine of data collection and ad targeting. Some of Kaplan's employees said the former Marine Corps artillery officer approached his job as a variation on tactical warfare. He followed a U.S. Marines leadership principle known as "JJDIDTIEBUCKLE," a fourteen-letter shorthand for qualities like justice, judgment, dependability, and initiative. It didn't matter so much who was in power, and it was a mistake to think of Kaplan as a political ideologue, some colleagues said. If he was at fault for following any ideology, it was the dogma of Beltway political insider-ism. His staff was almost equally divided between Republicans and Democrats. "It was business. He would do the same for a Democrat in the White House if he was a Democrat," one employee on the policy team explained.

Over the summer, Kaplan had begun to organize dinners and meetings for Zuckerberg with influential conservatives, including Sen. Lindsey Graham of South Carolina, a key Trump ally, and Tucker Carlson, the inflammatory Fox News host who had described white supremacy as a "hoax." Kaplan directed his Democratic lobbyists to arrange similar dinners with Facebook's biggest Democratic critics. The evening before the September Oval Office meeting, Zuckerberg met with Sens. Mark Warner of Virginia and Richard Blumenthal of Connecticut and three other lawmakers at the downtown restaurant Ris DC. Over grilled salmon and roasted brussels sprouts, Zuckerberg updated them on foreign accounts Facebook had taken down for spreading disinformation.

But employees on the West Coast viewed Kaplan's tactics as deeply cynical; Trump wasn't just another politician. They wanted Facebook to draw a bright line with his administration. And some thought Kaplan's protection of the business was shortsighted. As one mid-level MPK employee explained, "If you really want to protect the business, you have to protect users—their safety and their interests, not just the profits."

Trump was gearing up for a run at a second term using the same social media tactics as he did in 2016. On the other side of the Potomac River, a few miles from the White House, his campaign had set up a well-funded media operation, with Facebook again at the center of its strategy. From the fourteenth floor of an office tower in Rosslyn, Virginia, that used to house a trading company, Brad Parscale, Trump's campaign manager, had hired a phalanx of social media and campaign experts. They were planning to spend at least $100 million on Facebook ads, more than double the amount from the 2016 campaign.

The Trump campaign team's strategy was to overwhelm voters with posts by their candidate and Facebook ads. They created numerous iterations of a single ad, changing colors and tweaking a word here and there to appeal to specific demographics and voters in microtargeted geographies. The Facebook ads were often responses to negative narratives about Trump in the mainstream media. His actual political rival was less important, Trump's former senior adviser Steve Bannon had explained in a *Bloomberg* interview in February 2018. "The Democrats don't matter. The real opposition is the media," he said. "And the way to deal with them is to flood the zone with shit."

Sandberg wasn't taking part in Kaplan's dinners. Her focus was on improving relations with other critics. In September 2019, one week after Zuckerberg's Oval Office détente with Trump, she traveled to Atlanta as the featured speaker at the Civil Rights X Tech town hall, a daylong conference addressing discrimination in technology.

For years, civil rights leaders had accused Facebook of bias in its ad targeting and had scolded the company for barely increasing its percentage of Black and Hispanic employees from the low single digits of the total workforce. The 2016 election had exposed how the platform was also allowing Russian agents to exploit causes like Black Lives Matter, targeting Black voters. Hate was flourishing on the site. Employees of all races had become more vocal about discrimination inside the company and critical of its mostly white leadership.

Sandberg told aides it was one of the hardest issues for her, particularly accusations of anti-Semitism rising on the site. In May 2018, after pressure from human rights and civil rights leaders, she had begun a civil rights audit of the company and recruited Laura Murphy, a well-known civil rights leader formerly of the ACLU, to lead the process. She had also put Gene Sperling, the former director of the National Economic Council under Obama and a member of Sandberg's kitchen cabinet of advisers, in contact with Murphy and outside civil rights leaders to aid in their inquiry into the company and to come up with policies to protect the upcoming Census count and voter integrity. On June 30, 2019, she had released a clear-eyed update on the audit of the company's civil rights record, which confirmed a history of discrimination in housing and other advertising.

But the company was beginning to address some of its problems. In March 2019, Facebook had settled lawsuits against it for discriminatory targeting for housing, employment, and financial advertisements. Facebook implemented a new policy that banned age, gender, and zip code targeting for housing, jobs, and banking ads. That same month, the platform had announced a ban on white supremacist groups and speech that praised white nationalism. Sandberg announced the creation of a task force on civil rights and diversity, and she directed resources into a program to prevent misinformation around the Census 2020 count.

"Since I initiated my audit of Facebook, I've observed a greater willingness for Facebook to listen, adapt, and intervene when it comes to civil rights," Murphy concluded in her report. "I do believe the company is in a different place today than it was a year ago on these issues—I've seen employees asking the right questions, thinking about things through a civil rights lens, and poking holes in things on the front end of products and policies."

The Atlanta conference, which Sandberg attended with her son and mother, was meant to be the culmination of a hard-fought year of slowly repairing relations with civil rights leaders, with whom she had been consulting closely. "We appreciated Sheryl's efforts and invited her onstage to show we wanted to keep working together," recalled Rashad Robinson, the president of the civil rights group Color of Change and an organizer of the conference.

But two days before the Atlanta event, Clegg dropped a bombshell while speaking on a panel at the *Atlantic* Festival in Washington, DC: Facebook would give political leaders complete and unchecked voice. He was clarifying a murky "newsworthy" exemption that until then had been used only a few times. It was rooted in the decision to preserve Trump's Muslim ban video. It was formalized with the September 2016 decision to reinstate the iconic "Terror of War" Associated Press photo of a naked nine-year-old girl fleeing a napalm attack on her village during the Vietnam War. Facebook had initially taken the photo down because it violated the company's policy against child exploitation. Clegg's purpose that day was to declare publicly that the company had decided to make its seldom-used exemption a standard across the board for political speech, except for content that could lead to violence and other harms.

"Our role is to make sure that there is a level playing field but not to be a political participant ourselves," Clegg proclaimed in his plummy Oxbridge accent. Peering at the audience, he slowed down to emphasize the gravity of his words. "That's why I want to be really clear with you today—we do not submit speech by politicians to our independent fact-checkers, and we generally allow it on the platform even when it would otherwise breach our normal content rules," he added.

Political speech, Facebook spokespeople confirmed later, included paid ads by candidates and their campaigns. Clegg was confirming for the first time that political ads weren't fact-checked, which allowed politicians and their campaigns to pay to place lies on the site.

When Sandberg took the stage in Atlanta, the town hall audience was buzzing angrily about Clegg's speech. The policy he had described two days earlier was dangerous and could usher in a wave of misinformation and lead to voter suppression, some people in attendance said. American history was rife with examples of how, since the late 1800s, U.S. government leaders used Jim Crow laws, literacy tests, and poll taxes to reduce African American votes. "My colleagues and I arrived prepared to discuss solutions—not to fight new and flagrantly harmful policies," said Vanita Gupta, the president of the Leadership Conference on Civil and Human Rights. "We had traveled to Atlanta with confidence and left disturbed that the Facebook exemption would unravel our progress."

The Facebook officials at the event tried to defuse anger, insisting that the policy wasn't new and that the media were blowing Clegg's speech out of proportion. "That was spin," one civil rights leader said. "It was clear it was a new policy because we hadn't heard it before."

The timing couldn't have been worse for Sandberg. When she took the stage, she acknowledged the crowd's disappointment. "So, what I can't give you right now is perfect answers," she said, throwing her hands in the air. "We don't have them." But she wasn't deserting them, she added. "What you have from me, today and always, is my commitment to stay personally engaged, to stay close to you, and to make sure, not just me, but my company is listening."

Color of Change's Robinson had hired a crew of videographers to capture the Atlanta conference, which he intended to promote on his nonprofit's site. The morning of the event, he called off the video plans. "At the end of the day, the message was that Facebook is about business, and this is the way Facebook does business," he said.

To some civil rights leaders, Sandberg appeared blindsided by Clegg's announcement. According to one person close to the COO who attended the event, Sandberg claimed that so much had been going on at the time that she might have missed the details, even if they had been flagged for her. But Facebook officials said Sandberg was aware of Clegg's speech and that she agreed with the policy.

Within twenty-four hours of the Atlanta event, a grainy thirty-second attack ad against Democratic presidential candidate and former vice president Joseph Biden surfaced on Facebook. The ad, funded by a pro-Trump super PAC, started with a staticky video of Biden meeting with Ukrainian officials during his time in the Obama administration. "Joe Biden promised Ukraine $1 billion if they fired the prosecutor investigating his son's company," an ominous voiceover claimed; the video then cut to a Council on Foreign Relations event in which Mr. Biden referenced the sum. "But when President Trump asks Ukraine to investigate corruption, the Democrats want to impeach him," it read, a misleading allusion to the president's infamous July 25 phone call to newly elected Ukrainian president Volodymyr Zelensky in which Trump promised to release funds earmarked by Congress for Ukraine's war with Russia if Zelensky opened an investigation of Joe Biden and his son Hunter. The Trump campaign was misconstruing the events. There was no evidence that Biden had pushed for a dismissal of a Ukrainian prosecutor to help his son.

The ad spread for days on the social network, where it racked up more than five million views. On October 4, the Biden campaign wrote a letter to Zuckerberg and Sandberg with a demand that it be taken down. Neither Zuckerberg nor Sandberg replied.

On October 7, Katie Harbath, head of Facebook's engagement with political campaigns, wrote back. "Political speech is already arguably the most scrutinized speech there is," she noted, adding that the misleading ad would not be fact-checked and would remain on the site.

On October 17, Zuckerberg appeared at Georgetown University's campus in Washington, DC, to deliver his first major public address on Facebook's responsibility as a platform for speech. He came to defend his new policy for political misinformation, responding with defiance to criticism from the Biden campaign and scores of lawmakers.

He began his speech with a lie.

Zuckerberg stood behind a wooden lectern before 740 students, faculty, and media in the ornate Gaston Hall, a storied venue where presidents, religious leaders, and English royalty had spoken. Century-old murals of Jesuit figures and mythical Greek heroes adorned the walls. Zuckerberg's audience was quiet, and the mood was solemn. They applauded politely and stayed in their seats. It was nothing like when Sen. Bernie Sanders or President Obama spoke on campus. Then, students stood, and their cheers could be heard from outside the building.

Zuckerberg, with the sleeves of his black sweater pushed up on his forearms, began to read from the teleprompters. He offered a revisionist history of Facebook's genesis. While he was at Harvard, he told the audience, the United States had just gone to war in Iraq:

"The mood on our campus was disbelief," he said, adding that he and other students were eager to hear more perspectives about the war. "Those early years shaped my belief that giving everyone a voice empowers the powerless and pushes society to be better over time."

On Twitter, journalists and academics called out Zuckerberg for attempting to rewrite Facebook's origin story. The platform was widely understood to have begun with a project to rate hot Harvard girls, not quite the serious-minded undertaking he was describing fifteen years later on the Georgetown stage. And while protests had broken out at Harvard and other schools across the United States after the invasion of Iraq in 2003, Zuckerberg had not discussed, nor had he exhibited any interest in, the war, and his site was used for more mundane interests, friends recalled. "I was very surprised to hear that version," Chris Hughes said. "People largely used [Thefacebook] to share movie and music recommendations— not to protest foreign wars."

The story was a reminder of how much time had passed since the world was first introduced to Facebook. Once viewed as a hero hacker among college students, Zuckerberg now came off as a rich thirty-five-year-old father of two. The college students were almost a full generation younger than he. They weren't using Facebook, a site popular with older audiences. Many were on Instagram but were increasingly spending time on Snapchat and TikTok.

Georgetown was a deliberate choice. Facebook's policy and lobbying staff wanted to find a place for Zuckerberg to deliver his speech where his words would carry intellectual and historical import. The staff wanted to do it in Washington, with Zuckerberg's most important viewers a few miles east, at the White House and on Capitol Hill. Facebook staff treated it like a major political event, sending an advance team to sweep the hall and plan out the choreography for Zuckerberg's entrance. Next to the wooden lectern at the center of the stage, they arranged for teleprompters. And the Facebook CEO had worked on the speech for weeks.

Zuckerberg then described the importance of free speech in the civil rights movement. With every social movement comes a clampdown on free expression. He evoked the Black Lives Matter movement, Frederick Douglass, and Martin Luther King Jr., who was jailed for protesting peacefully. "In times of social tension, our impulse is often to pull back on free expression. Because we want the progress that comes from free expression, but we don't want the tension," he said.

The internet had created a powerful new form of free expression, he continued. It had removed the "gatekeepers" of traditional news organizations, giving any individual a voice that could be amplified across the globe. Unlike the "fourth estate," the media that held kings and presidents to account, Facebook was part of a new force that he described as "the fifth estate," which provided an unfiltered and unedited voice to its 2.7 billion users. He warned against shutting down dissenting views. The cacophony of voices would of course be discomfiting, but debate was essential to a healthy democracy. The public would act as the fact-checkers of a politician's lies. It wasn't the role of a business to make such consequential governance decisions, he said.

He ended his speech without taking questions.

The blowback was immediate. "I heard #MarkZuckerberg's 'free expression' speech, in which he referenced my father," Martin Luther King Jr.'s daughter Bernice King tweeted. "I'd like to help Facebook better understand the challenges #MLK faced from disinformation campaigns launched by politicians. These campaigns created an atmosphere for his assassination."

Zuckerberg hadn't been involved in the company's civil rights work. When Vanita Gupta, who had been pushing him for changes on the platform to combat disinformation, asked him during a phone conversation days before the Georgetown speech if he had civil rights experts on staff, he responded that he had hired people from the Obama administration. He seemed to be implying that human rights and civil rights were Democratic causes. In his concern for balancing Republican and Democratic issues, civil rights fell on his list of liberal causes. "Civil rights are not partisan. It was troubling to hear that," Gupta said.

Lawmakers, consumer groups, and civil rights advocates warned that they would have to fight Facebook to protect the 2020 elections. Sherrilyn Ifill, president of the NAACP's Legal Defense Fund and a renowned expert on voting rights, cautioned that Zuckerberg had displayed a "dangerous misunderstanding of the political and digital landscape we now inhabit." The civil rights movement Zuckerberg had evoked fought first and foremost to protect citizenship and human rights guaranteed to Black people under the Fourteenth Amendment. It was not primarily a movement to protect First Amendment rights, she pointed out. Zuckerberg had also misinterpreted the First Amendment, which was designed to protect speech from government censorship. The very people Zuckerberg aimed to protect, political figures, could wield the most harm: "The company has refused to fully recognize the threat of voter suppression and intimidation here at home, especially from users that the company refers to as 'authentic voices'—politicians and candidates for public office," Ifill asserted.

Academics viewed the policy as disingenuous. The right to free expression didn't include the right to "algorithmic amplification," said Renée DiResta, who had become a widely recognized expert on social media companies and misinformation online from her work on Russian election interference. The technology created echo chambers of like-minded individuals sharing the same stories. Zuckerberg was having it both ways. Facebook had become as powerful as a nation-state, bigger than the most populous country in the world. But nations were governed by laws, and their leaders invested in public services like firefighters and police to protect their citizens. Zuckerberg wasn't taking responsibility to protect Facebook's members.

Others at the company agreed with DiResta. "Free speech and paid speech are not the same thing," read an open letter signed by hundreds of Facebook employees and addressed to Zuckerberg. "Misinformation affects us all. Our current policies on fact checking people in political office, or those running for office, are a threat to what FB stands for. We strongly object to this policy as it stands. It doesn't protect voices, but instead allows politicians to weaponize our platform by targeting people who believe that content posted by political figures is trustworthy."

One week later, Zuckerberg was back in Washington for a hearing before the House Financial Services Committee about his pet project, Libra. He saw the blockchain currency as eventually replacing currency systems already in place, putting Facebook at the forefront of monetary policy and systems. During his trip, he met Trump again, this time for a secret dinner at the White House. He brought Priscilla, and they were joined by First Lady Melania Trump, Jared Kushner, Ivanka Trump, Peter Thiel, and Thiel's husband, Matt Danzeisen. It was a friendly dinner meant to acquaint the couples. The White House was warming to Zuckerberg, who appeared to be showing courage in ignoring pressure from Facebook's critics.

California Democrat Maxine Waters, the chair of the committee, began the hearing the next day with a call to stop the blockchain currency project, which she claimed was riddled with privacy, security, discriminatory, and criminal risks. Facebook had way too many problems already; Zuckerberg's insistence on proceeding with Libra, she said, reflected his company's cutthroat culture. Zuckerberg was willing to "step on or over anyone, including your competitors, women, people of color, your own users, and even our democracy to get what you want." For the next five hours, lawmakers unloaded on Zuckerberg over his leadership and his decision to host political lies on his platform. If any Facebook allies were present, they didn't speak up to defend him.

Democratic freshman representative Alexandria Ocasio-Cortez of New York asked how far she could go in spreading lies on Facebook as a politician. "Could I run ads targeting Republicans in primaries saying that they voted for the Green New Deal?" she asked. "Do you see a potential problem here with a complete lack of fact-checking on political advertisements?" Zuckerberg responded that he thought lies were "bad." But, he added, he believed "that people should be able to see for themselves what politicians that they may or may not vote for are saying."

Joyce Beatty, a Democrat from Ohio, grilled him about Facebook's civil rights abuses. He couldn't answer her question on how many African Americans used the site, a figure widely publicized by Pew Research. She quizzed him on the recommendations contained in Facebook's own civil rights audit. He stuttered and took a big gulp from a plastic water bottle. When Zuckerberg said one recommendation was to form a task force that included Sandberg, Beatty scoffed. "We know Sheryl's not really civil rights," she said and chuckled dismissively, adding, "I don't think there's anything—and I know Sheryl well—about civil rights in her background." She accused Zuckerberg of treating discrimination and civil rights abuses as trivial concerns. "This is appalling and disgusting to me," Beatty said.

It seemed the public had soured on Zuckerberg as well. A survey conducted by the Factual Democracy Project in March 2018 had found that just 24 percent held a favorable opinion of the CEO.

In an interview with Lester Holt of NBC News later that week,

Zuckerberg reflected on the evolution of his fifteen-year leadership. "For a long time, I had a lot of advisers . . . who told me when I go out and communicate, I should focus on not offending anyone," he said. "Part of growing up for me has just been realizing that it is more important to be understood than it is to be liked."

Public appearances had become increasingly difficult for Sandberg, but *Vanity Fair*'s New Establishment Summit felt like a safe choice. The event, which was billed as bringing together "titans of technology, politics, business, media, and the arts for inspiring conversations on the issues and innovations shaping the future," was seen as more of a lavish networking gathering than a hard-hitting news event. The two-day program, which ran from October 21 to 23, 2019, featured speakers like actress and lifestyle entrepreneur Gwyneth Paltrow, Disney CEO Bob Iger, and former Trump communications director Anthony Scaramucci. Adding to the comfort level was the fact that Sandberg was paired with a longtime friend and confidante for the interview, the veteran broadcast journalist Katie Couric. The two women had bonded over their shared experience of being widowed young; Couric had lost her husband to colon cancer at age forty-two and had supported Sandberg's book *Option B* with interviews at public events.

The two women spent time in the greenroom catching up before the event. Couric knew Sandberg's new boyfriend, Tom Bernthal. Before he started a private consultancy firm, Bernthal had worked as a news producer at NBC, and his time there overlapped with Couric's, who spent fifteen years on air as a star presenter and anchor. When Sandberg and Couric got onstage, they continued their easy rapport, poking fun at their promotional cartoon avatars beamed onto a large screen high up behind their chairs.

But once the interview began, the tenor changed. Wouldn't Zuckerberg's new policy on political ads directly undermine efforts to counter election interference? Couric asked. "The Rand Corporation actually has a term for this, which is 'truth decay.' And Mark himself has defended this decision even as he expressed concern about the erosion of truth online," she added. "What is the rationale for that?"

For nearly fifty minutes, the former news anchor pressed Sandberg with a wide range of tough and intense questions. She asked about the critical response from civil rights leaders to Zuckerberg's speech at Georgetown. She asked about bullying on Instagram and Facebook. She grilled Sandberg on the echo chambers Facebook created and the algorithms that worked against more centrist politicians like Joe Biden. She pushed the COO to defend Facebook against Senator Warren's call to break up the company and asked skeptically if the reforms Sandberg had promised would be effective against Facebook's business model of viral ads.

Sandberg stuttered through some responses and gave template answers. Several times, she conceded that the issues were difficult and that Facebook felt responsible, but she stopped short of saying that the company would fact-check speech. She noted that she knew Bernice King personally and had spoken to her about her tweet criticizing Zuckerberg's Georgetown speech. She pointed to the disagreement between King and Zuckerberg as an example of important civil discourse, underscoring Facebook's belief in free expression and the clash of opposing ideas. Attempting to bring some levity back to the discussion, she added that she wished King had posted her thoughts on Facebook instead of Twitter. The joke was met with silence from the audience.

Toward the end of the conversation, Couric posed the question that few were bold enough to ask Sandberg directly: "Since you are so associated with Facebook, how worried are you about your personal legacy as a result of your association with this company?"

Sandberg didn't skip a beat as she reverted to the message she had delivered from her first days at Facebook. Inside, she was burning with humiliation, she later told aides, but she kept her voice steady and a smile on her face. "I really believe in what I said about people having voice. There are a lot of problems to fix. They are real, and I have a real responsibility to do it. I feel honored to do it."

The truth was more complicated. To friends and peers, Sandberg tried to dissociate herself from Zuckerberg's positions, but within the company, she executed his decisions. After Zuckerberg delivered his speech at Georgetown, the Anti-Defamation League admonished him for emboldening racists and anti-Semites on Facebook. Civil rights groups, including the ADL, warned that his new policy was being introduced just as hate speech was exploding on the site. Even generous charitable donations could not erase the damage Facebook was wreaking.

There was little she could do to change Zuckerberg's mind, Sandberg confided to those close to her. When an outside consultant fired off a series of angry emails about Zuckerberg's Georgetown speech, Sandberg wrote back that he should forward the emails to Clegg and others who might influence Zuckerberg's thinking. Her inaction infuriated colleagues and some of her lieutenants—his decisions, after all, were in direct contradiction to the core values she promoted in public.

Sandberg justified her inaction by pointing to Mark's powerful grip as Facebook's leader. "She was very careful about picking her battles with him," one former employee said. She was second-in-command in title, but the top ranks were crowded: Clegg, Kaplan, and executives like Olivan were taking on bigger roles, alongside Adam Mosseri, the head of Instagram, and Guy Rosen, the former cofounder of Onavo and now head of integrity at Facebook.

Clegg was handling more television interviews, which Sandberg was happy to hand off, her aides said.

She had become less assured. Media coverage of her was hard-hitting and focused on her leadership failures. She was receding into the background, while also growing defensive. "Her desire to appear perfect gets in her way of defending herself," a friend said.

An organization Sandberg had championed for years was also escalating an offensive against Facebook. In November, the British comedian Sacha Baron Cohen was scheduled to deliver a speech at an ADL summit. Baron Cohen planned to speak out about the hate speech and anti-Semitism flooding social media. He reached out to Common Sense Media's Steyer and others in the "Coalition of the Willing," the group assembled by law professors Tim Wu and Scott Hemphill, to help with his speech.

On November 21, Baron Cohen delivered a twenty-four-minute rebuke of "the Silicon Six" for spreading hate. He dismissed Zuckerberg's Georgetown speech as "simply bullshit": "This is not about limiting anyone's free speech," he said. "This is about giving people, including some of the most reprehensible people on earth, the biggest platform in history to reach a third of the planet."

Every January, for more than a decade, Zuckerberg had posted a personal resolution for the year ahead. In 2010, he set out to learn Mandarin. In 2015, he resolved to read a book every two weeks. In 2016, he said he would program AI for his home and run a mile a day.

Starting in 2018, his resolutions carried greater urgency and weight. That year, he had pledged to focus on fixing Facebook's myriad problems: "protecting our community from abuse and hate, defending against interference by nation states, or making sure that time spent on Facebook is time well spent." In 2019, he announced that he would host a series of discussions with thought leaders on the role of technology in society and the future.

For 2020, he announced a more audacious goal. Instead of an annual challenge, Zuckerberg would tackle some of the most intractable problems facing the globe over the next decade. In a 1,500-word post, he reflected on his role as a father and on his nonprofit work to cure disease and extend life expectancy. He was presenting himself as an elder statesman, like his mentor Bill Gates, who had been counseling him on philanthropy and in whose footsteps he hoped to follow, pivoting from tech CEO to global philanthropist. Zuckerberg had long been exasperated by the lack of credit and attention he received for his charitable efforts, he told staff and pollsters who tracked his reputation in regular public surveys. His philanthropy with wife Priscilla, the Chan Zuckerberg Initiative, was a case in point. It had been announced in December 2015 with a big media rollout. And yet initial praise for the foundation, to which he dedicated the majority of his wealth, was overshadowed by skeptical news reports that fixated on the decision to set it up as a limited liability corporation, noting that this designation provided a tax haven for the young billionaire couple. Buffeted by several days of negative stories, Zuckerberg called Sandberg and Elliot Schrage, who had helped strategize the press rollout, from his vacation home in Hawaii, screaming that they had screwed up. It was an uncharacteristic display of anger that reflected Zuckerberg's frustration with his difficulty at winning over public goodwill. "He was very mad about the press around CZI," one former employee recalled. "He would say, 'Why don't people think of me the same way as Bill Gates?'"

On January 31, 2020, Zuckerberg appeared at the Silicon Slopes Tech Summit, a tech conference held in Salt Lake City. He admitted that he was a poor communicator; for most of Facebook's history, he said, he had been able to avoid putting his intentions front and center. But Facebook had become too important, and as its leader, he was compelled to stand behind his positions publicly. His company was devoted to the absolute ideal of free expression, he explained. "The line needs to be held at some point," he added. "This is the new approach, and I think it's going to piss off a lot of people. But, you know, frankly, the old approach was pissing off a lot of people, too, so let's try something different."

When asked if he felt like he was taking heat on behalf of "the entire internet," Zuckerberg broke into laughter. It took him a few seconds to collect himself. "Um. That's what leading is," he said.

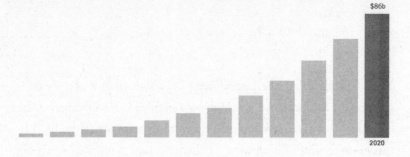

$86b

2020

Chapter 14

Good for the World

Well before U.S. government officials and much of the world were aware, Zuckerberg knew that the COVID-19 virus was spreading dangerously fast. In mid-January, he began to receive intelligence briefings on the global spread of the disease from top infectious disease experts working at the Chan Zuckerberg Initiative. Dr. Tom Frieden, the former head of the Centers for Disease Control and Prevention, and Joe DeRisi, co-president of the Biohub funded by CZI, reported that the Chinese government and Trump were downplaying the risk and that the virus had spread to half a dozen countries. Facebook had the potential to play a critical role in what appeared to be the first global pandemic of the internet age.

On January 26, Zuckerberg ordered his department heads to drop all nonessential work to prepare. The civic engagement team, which had been developing a number of new features ahead of the presidential election, including a centralized hub for sharing voting information, shifted its efforts to create a center for information from the CDC and the World Health Organization. Fact-checking tools to label misinformation would be used to correct conspiracy theories about COVID-19. Facebook's HR teams would begin drafting a work-from-home policy as Facebook's offices across Asia were tasked with gathering information about what was truly happening in China. And Zuckerberg's personal relationship to the world's most famous doctors and health experts would be used to promote figures of authority. Zuckerberg told the teams to report back to him in forty-eight hours. "This is going to be big, and we have to be ready," he said.

When the WHO designated COVID-19 a public health emergency on January 30, Facebook's PR team had a blog post prepared. Within hours, the company went public with its plan to remove harmful misinformation, connect people to authoritative information on the virus, and give unlimited ad credits to the WHO and the CDC to run public service ads about the novel coronavirus. Facebook was the first company in Silicon Valley to issue a response.

Zuckerberg was also one of the first American CEOs to shut down offices and authorize employees to work from home. He sent a memo to employees detailing how the coronavirus would affect every aspect of Facebook's business. The company would likely lose revenue from advertisements as the pandemic crippled businesses across the economy. Facebook's infrastructure, including its data centers, would be tested to the breaking point as billions of people were online at the same time.

Positive headlines about the company's response provided a boost to employee morale. Zuckerberg had an opportunity to burnish his reputation as a leader. Along with Priscilla, who had practiced medicine in San Francisco for years, he hosted Dr. Anthony Fauci on a Facebook Live interview watched by more than five million people, and he went on to record live videos with others, including California governor Gavin Newsom and Dr. Don Ganem, a virologist and infectious disease specialist. News shows began to invite Zuckerberg to speak about his views on the pandemic response.

One month into the pandemic, Zuckerberg was sent data that revealed that the PR blitz was making a difference to Facebook's public image. For years, the company had been running daily surveys in which it asked users a series of questions about their impressions of the social network. Zuckerberg tracked two questions in particular: whether respondents felt that Facebook was "good for the world" and whether they thought Facebook "cares about users." He cited stats on the questions so frequently that, in meetings, they were referred to by their respective initials, "GFW" and "CAU." In early April, the number of people who saw Facebook as "good for the world" began to show marked improvement for the first time since the Cambridge Analytica scandal.

But at a White House briefing on April 23, Trump put the company to the test when he suggested that disinfectants and ultraviolet light were possible treatments for the novel coronavirus. His remarks, which were quickly debunked by doctors and health professionals, went viral. Within hours of Trump's comments, more than five thousand posts on the topic sprang up on Facebook and Instagram, viewed by tens of millions of people. Facebook removed a few, but the source of the false and dangerous information, President Trump's Facebook account, remained untouched. Trump had, once again, proved to be the perfect foil for what Zuckerberg saw as his well-intended positions. For each of Zuckerberg's assumptions about how a political leader with Trump's following could, or would, use a platform like Facebook, the president defied it, bypassing Facebook's rules and attracting record engagement in the process.

For over a day, the comms team dodged questions from the press. Journalists pointed out that Zuckerberg had previously included medical misinformation among the few exceptions for free political speech. Wasn't this exactly the type of situation Facebook had said it would take action against?

Facebook's policy and content teams met to discuss how to proceed. Kaplan sided with his team, which argued that Trump had simply been *musing* about the effects of bleach and UV light, rather than issuing a *directive*. Facebook's PR teams pointed out the distinction to journalists and released a statement saying, "We continue to remove definitive claims about false cures for Covid-19, including ones related to disinfectant and ultraviolet light."

"We, all of us at Facebook, just couldn't get our heads around the idea that it would be the president who spread nonsense cures about COVID," said one senior Facebook executive. "We were just stuck in an untenable position. It meant defending a decision to let the president spread absolute nonsense medical information in the middle of a pandemic."

On May 29, 2020, Twitter took the unprecedented step of slapping a warning label on one of Trump's tweets.

At 1 a.m., the president had posted a message on his Facebook and Twitter accounts about the protests roiling American cities. Americans had taken to the streets in response to the death of George Floyd, a forty-six-year-old Black man whose neck was pinned under the knee of a Minneapolis police officer for over nine minutes for the crime of purchasing a pack of cigarettes with a counterfeit twenty-dollar bill. Trump wrote that he was concerned about the protests and was in contact with Minnesota governor Tim Walz, to offer the assistance of the U.S. military. "Any difficulty and we will assume control but, when the looting starts, the shooting starts," read one of his posts. Between his Facebook and Twitter accounts, Trump's words reached more than one hundred million Americans. Neither company had ever removed or labeled his posts, per a position they shared that the president was in a special class of protected speech.

The phrase "when the looting starts, the shooting starts" mirrored language used by a Miami police chief in the late 1960s following riots in that city. It delivered a twofold insult, likening thousands of peaceful protesters to looters, and suggesting that police should open fire on the crowds gathered under peaceful assembly. For Twitter, it was a step too far. Within hours, the platform had put a warning label on the tweet, indicating that Trump was glorifying violence. Trump fired back, threatening regulations and the abolishment of an important legal shield for internet companies known as Section 230. "Twitter is doing nothing about all of the lies & propaganda being put out by China or the Radical Left Democrat Party. They have targeted Republicans, Conservatives & the President of the United States. Section 230 should be revoked by Congress. Until then, it will be regulated!" he tweeted.

Over the past year, Twitter had been gradually diverging from Facebook on speech policies. It had banned political ads soon after Zuckerberg's Georgetown declaration to protect all political speech, and it had begun to label tweets in which Trump spread false or misleading information around the elections; earlier in the week, Twitter had attached a fact-checking label to two tweets by the president that discredited the integrity of mail-in voting. But the warning labels on the Black Lives Matter demonstrations tweets marked a shift: it was not a question of correcting a falsehood, but of calling out the potential danger of Trump's words. Twitter was clearly stating that it was aware that the president was using its platform to spread an idea or message that could lead to real-world harm.

The move by Twitter put more pressure on leaders at Facebook to take a stand against the president's dangerous rhetoric. The usual suspects met, once again, to discuss their options regarding Trump's account. The policy team, including Kaplan, reprised the argument used to defend Trump's bleach comments: there was no clear directive; Trump had not actually ordered anyone to open fire on protesters. The team sent its decision to Zuckerberg and told the White House that it was reviewing Trump's post and weighing options.

Later that day, Trump called Zuckerberg. The president wanted to plead his case and ensure that his Facebook account was safe. Zuckerberg explained that he was concerned about the divisive and inflammatory nature of the post. He told the president he was disappointed with his use of the "looting" and "shooting" rhetoric. "Zuckerberg mainly listened while Trump did his usual talking in circles," recalled one Facebook employee briefed on the conversation. "I wouldn't say Trump apologized for the post. He did the bare minimum so that Facebook wouldn't remove it." Zuckerberg attempted to frame a discussion around responsibility and the power of Trump's audience on Facebook, reminding the president that he had a personal responsibility to use his massive reach on the platform responsibly. But he also made it clear that he would not be removing the post or taking action against Trump's account.

What had begun as one-off decisions had matured into a policy that made politicians and public figures a special protected class on Facebook. Zuckerberg justified the position in a post on his Facebook page. He found Trump's comments personally offensive, he wrote. "But I'm responsible for reacting not just in my personal capacity but as the leader of an institution committed to free expression."

Zuckerberg's post was shared widely within the company, along with a clip from a Fox News interview earlier in the month in which he seemed to criticize Twitter's approach to Trump. "We have a different policy I think, than Twitter on this," Zuckerberg had said. "I just believe strongly that Facebook shouldn't be the arbiter of the truth of everything that people say online."

The contrast between the two companies was demoralizing to Facebook employees. On their Tribe boards, employees asked if there were job openings at Twitter. One engineer went a step farther, asking if there was a job at "any company which showed they were willing to take a stance on their moral responsibility in the world—because Facebook doesn't seem to be it." Another employee posted an internal poll asking his fellow Facebook employees if they agreed with Zuckerberg's decision. More than a thousand people responded that the company had made the wrong call.

That Monday, hundreds of employees staged a virtual walkout and changed their Workplace avatars to a black-and-white fist. Some also took the unprecedented step of posting public messages on their social media accounts and out-of-office email notices, criticizing Facebook directly for its decision. "The hateful rhetoric advocating violence against Black demonstrators by the U.S. President does not warrant defense under the guise of freedom of expression," Robert Traynham, a senior Black employee and director of policy communications, wrote on the Tribe group for Black@ Facebook. "Along with Black employees in the company, and all persons with a moral conscience, I am calling for Mark to immediately take down the President's post advocating violence, murder and imminent threat against Black people." Traynham's post built on years of frustration among Black employees, many of whom no longer trusted their employer enough to use the Black@Facebook Tribe board to air grievances about the company. They had taken to meeting in person or discussing problems through encrypted apps; now they banded together in support of the walkout and the pressure it would put on Facebook executives. "For years, the company had been trying to appease us with claims that they were working on diversity and were committed to racial justice. All of that means nothing when Mark is willing to give the president of the United States a free pass to target black people on Facebook," wrote another member of the Black@Facebook group. "Stop letting people be racist on Facebook, maybe then we will believe that you support racial justice."

Three days later, thirty-three former early employees signed and distributed to the media an open letter of protest, asserting that Facebook had become unrecognizable to them: "It is our shared heartbreak that motivates this letter. We are devastated to see something we built and something we believed would make the world a better place lose its way so profoundly. We understand it is hard to answer these questions at scale, but it was also hard to build the platform that created these problems." The letter ended with a direct appeal to Zuckerberg: "Please reconsider your position."

At his weekly Q&A meetings that month, Zuckerberg seemed visibly shaken. The unrest at his company was being leaked to the press at record speed. Reporters at various news outlets, including the *New York Times*, the *Washington Post*, and the Verge, had been publishing leaks from Q&As for years. But recently, a reporter at BuzzFeed, Ryan Mac, had been tweeting out notes from the weekly meeting in real time.

For eighty-five minutes straight at one meeting, Zuckerberg fielded questions about his decision to let Trump's post stand, sticking closely to his script about the importance of free speech. When he was asked whether he had consulted any Black employees in reaching his decision, he mentioned Maxine Williams, a member of Zuckerberg and Sandberg's inner circle who had served as Facebook's head of diversity since 2013. But few in Facebook's ranks had reason to trust Williams, under whose tenure little had improved. In 2014, Facebook had reported that 2 percent of its employees were Black; five years after the company promised to make diversity a priority, the number had increased to just 3.8 percent. And Williams didn't speak up for Black employees during the walkouts, several members of the Black@Facebook Tribe said. Zuckerberg's singling her out did not reassure anyone at the meeting. "If we had been there, in person, you would have heard the groans in the room," one employee recalled.

Zuckerberg then announced that Williams would report directly to Sandberg, a decision intended to underscore the company's commitment to diversity and to Black employees. But, to many inside and outside the company, the plan registered as a PR strategy. Some employees had lost faith in Sandberg, who from her earliest days had promised to improve diversity in hiring and who had been charged with addressing civil rights abuses on the platform. But her efforts, which included highly publicized audits and college recruitment efforts at Historically Black Colleges and Universities, weren't yielding meaningful improvements. Her role in the company had become less clear in general throughout the crisis years; many employees and outside observers felt that she had faded into the background. Nick Clegg and Zuckerberg were fielding many of the public speaking engagements and news interviews; Sandberg did not publicly comment when she disagreed with Zuckerberg's speech policies. "It was a huge disappointment," one former employee recalled. "We were waiting for her to say something about Trump, especially the 'looting and shooting' post, and she was silent. Either she has no power or she is complicit, or both."

As Zuckerberg worked alternately from his home in Palo Alto and his ocean-side ranch in Hawaii, he watched protests swelling across the nation and Trump ramping up his efforts to discredit the legitimacy of mail-in ballots in the upcoming election. In July 2020, Zuckerberg confided to Vanita Gupta of the Leadership Conference on Civil and Human Rights that he worried about polarization and disinformation and the effect of Trump's use of the platform. "I'm really afraid for our democracy," he told her in a phone call. Trump's posts were "undermining our policy and are pushing authoritarianism." The one person who benefited the most from Facebook's position on free expression was abusing the privilege. Zuckerberg asked Gupta, who had been frustrated with him for dragging his feet on election disinformation, for help in strengthening policy enforcement ahead of November. He said he would be personally involved: "I am feeling a great sense of urgency."

Part of the problem was that the most extreme behavior was taking part in closed or private groups. Facebook exacerbated the problem when, years earlier, it began its push to encourage people to join new groups. Research had shown that people who joined many groups were more likely to spend more time on Facebook, and Zuckerberg had hailed groups as the type of private, living room chat he thought his users wanted to see more of. But he was growing disturbed by the number of people joining groups dedicated to conspiracy theories or fringe political movements, rather than the hiking clubs and parenting communities he had envisioned. "He seemed, for the first time, actually personally and philosophically disturbed," Gupta said.

It no doubt helped that there was a voice in Zuckerberg's ear contradicting Kaplan's these days. Cox, who had returned to Facebook as chief product officer the previous month, had been raising the topic of speech with Zuckerberg directly in conversation. During his year away from the company, Cox advised a Democratic political nonprofit to get Trump voted out of office, and he had spoken openly about his belief that toxic content should not be allowed to go viral and of the need to fact-check political ads. "Social media companies and messaging companies should take responsibility for things that get lots of distribution," he had declaimed not long after his resignation, in a talk to South Park Commons, a members-only tech community founded by his News Feed co-creator, Ruchi Sanghvi. Cox had spent a lot of time in meditation during his time away from Facebook and had undergone a radical evolution in thinking from his early engineering years, when he saw criticism of News Feed and other products as overblown. "I'm finding these problems of misinformation and privacy, sort of the ability to aggregate a bunch of data, as being really problematic and scary because they create asymmetric concentration of power which can be abused by bad actors."

It was widely known that Zuckerberg and Cox had remained close in the wake of the CPO's departure, and so it did not come as a surprise to anyone at Facebook when the two quietly negotiated Cox's return to the executive ranks. But it also spoke volumes that Zuckerberg had brought Cox back despite the criticism he had leveled at the platform. Indeed, some employees wondered if he was asked to return because of his conflicting perspective. "Mark trusted Cox to offer an outside point of view, because he knew that Cox ultimately wanted Facebook to succeed as much as he did. Cox was one of the only people he could wholly trust in that way," said one Facebook executive. "Having him back at Facebook was a huge help to Zuckerberg, actually. It brought back his main sounding board."

For his part, as one of the company's earliest engineers, Cox felt a responsibility to try to stem damage from the technology he had helped create. "Facebook and our products have never been more relevant to our future. It's the place I know best, it's a place I've helped to build, and it's the best place for me to roll up my sleeves and dig in to help," he wrote in a Facebook post explaining his return.

He was quickly ushered back into Zuckerberg's inner circle of executives and jumped into the fray of Facebook's most fraught policy discussions. On July 7, Zuckerberg, Sandberg, Clegg, and Cox held a videoconference with the organizers of Stop Hate for Profit, an initiative spearheaded by a broad group of critics, including Jim Steyer of Common Sense, Sacha Baron Cohen, and civil rights leaders who had tapped their networks of business executives to participate in a one-month ad boycott of Facebook. By early July, Verizon, Starbucks, Ford, and Walgreens had announced that they were joining the protest. Sandberg had frantically called the CEOs and marketing chiefs of the more than one hundred participating companies, to try to get them to change their minds, to no avail.

The civil rights leaders on the call reprimanded the Facebook executives for the rise in hate speech on the site. Jonathan Greenblatt of the ADL interrogated Zuckerberg for allowing Holocaust deniers to continue to spread their views on the site. Rashad Robinson of Color of Change insisted that the company at least follow Twitter's lead in policing Trump's account for speech that incited violence and spread disinformation about mail-in voting.

Zuckerberg fielded most of the questions. He defended Facebook's efforts, saying the platform's AI was now able to detect 90 percent of hate and other harmful speech. It was a huge improvement and a testament to Facebook's artificial intelligence capabilities and its larger staff of content moderators. But with hundreds of millions of pieces of content produced each day, the 10 percent that remained to be found through manual or other means still represented millions of posts filled with hate speech. Greenblatt retorted that Zuckerberg's math was dangerous: you would never say that a coffee made up of only 10 percent carcinogens was safe.

Robinson took aim at Kaplan, who was not on the call. "We could keep pushing for better policies, but you aren't going to enforce your policies because you have an incentive structure rooted in keeping Trump not mad and avoiding regulation," he pointed out. "Your DC office and Joel are making these decisions."

Robinson's directness appeared to irk Zuckerberg. He didn't like having his back up against the wall. He also felt protective of Kaplan, who had been cast as a villain in numerous press stories, portrayed as a political mercenary devoted to empowering Trump on the site. "You're looking for a gotcha moment," Zuckerberg said, according to Robinson, his voice rising. "But I make the decisions here."

Sandberg, Cox, and others had been pressing Zuckerberg to reconsider his position on the Holocaust material. There was merit to the concerns raised by the ADL and other civil rights groups; the company needed to do more to clamp down on hateful and toxic speech. Facebook's enormous size and its technology were amplifying voices in the fringe right wing across its networks at unprecedented scale. Zuckerberg had defended the content as a matter of principle, but it was no longer a fringe problem, and it needed to be reassessed.

Zuckerberg was also seeing disturbing new data. One internal report revealed that Holocaust conspiracy theories and denialism were growing on Facebook and that Millennials in significant numbers were adopting these ideas. Some users believed that details of the Holocaust had been exaggerated; others believed that the entire event, which saw six million Jews targeted and systematically killed by Nazi Germany and its allies, had been fabricated. Zuckerberg was particularly struck by the detail about Millennials. It was incomprehensible to him that his own generation, raised with all the access to information the internet held, could suddenly turn around and deny the Holocaust. Sandberg and others pressed him to ask himself if it was a position he could continue to defend, especially in light of growing anti-Semitic attacks in the United States and elsewhere.

In late summer 2020, Zuckerberg ordered Facebook's policy team to begin crafting a policy to ban Holocaust deniers from the platform. "It was a total reversal, but Zuckerberg weirdly didn't see it like that. He talked about it as an 'evolution,'" said one person on Facebook's communications team who had been briefed on how to explain Zuckerberg's decision to the media. "We were announcing all these changes to long-standing policy but treating them as ad hoc, isolated decisions."

Other data were equally alarming. Internal reports also showed a steady rise in extremist groups and conspiracy movements. Facebook's security team reported incidents of real-world violence, as well as frightening comments made in private groups. Facebook's data scientists and security officials noted a 300 percent increase, from June through August 2020, in content related to the conspiracy theory QAnon. QAnon believers perpetuated a false theory that liberal elites and celebrities like Bill Gates, Tom Hanks, and George Soros ran a global child trafficking ring. They built their following on the foundation laid by "Pizzagate," a conspiracy theory that claimed that Hillary Clinton and other prominent Democrats were abusing children in the basement of a Washington, DC, restaurant. While the theory was repeatedly proven false—the restaurant in question did not even have a basement—the idea that there was a conspiracy being hatched among the global elites persisted and grew under the Trump administration.

On August 19, Zuckerberg agreed to take down some QAnon content on the grounds that it could lead to violence. It was a narrow change in policy, pertaining to a small portion of all QAnon content. And in a nod to what some team members called Kaplan's relentless internal lobbying for political equivalency, Facebook also announced that it would remove 980 groups, such as those related to the far-left antifa movement, a bugaboo for Trump and Republicans, who blamed violent demonstrations on leftist groups. "I can't explain it. We are not saying they are both the same type of violent group. All I can tell you is that we had to announce both of them the same day," said one Facebook engineer on an exasperated call with a reporter the day of Facebook's announcement. "It's political, okay? We can't announce QAnon without announcing something on the left."

The Facebook team dedicated to finding and removing extremist content on the site felt Zuckerberg's decision was a good first step, but they were closely monitoring how QAnon and other right-wing groups responded. They knew that as the election neared, the potential for violence would become more acute. On the morning of August 25, their fears were realized as escalating right-wing anger on the site boiled over. At 10:44 a.m., an administrator of a Facebook page named "Kenosha Guard" called out to members across Ohio and neighboring states to "take up arms and defend out [*sic*] City tonight from the evil thugs." That evening, an event called "Armed Citizens to Protect Our Lives and Property" posted. More than 300 users marked that they would attend, and over 2,300 expressed interest in the event. "I fully plan to kill looters and rioters tonight," one user posted in a comment on the page. "Now its [*sic*] the time to switch to real bullets, and put a stop to these impetuous children rioting," another member of the page commented. Others included details on what type of guns (the AR-15 rifle) and what type of bullets (hollow-point, expanding upon impact) would inflict the most damage.

Over the course of the day, Facebook received 455 complaints about the page highlighting specific comments that violated Facebook's rules on invoking violence. But as the hour of the protest neared, the page remained online.

Clashes between protesters and a group of armed men who identified as the "Kenosha Guard" started at nightfall. At 11:45 p.m. local time in Kenosha, two people were shot and killed and a third injured by a young man allegedly carrying an AR-15-style rifle.

Zuckerberg was notified that night about the shootings and the possible link to Facebook. He asked his policy teams to investigate. Six days earlier, the platform had passed a policy banning exactly these types of extremist groups and militias. Why, asked Zuckerberg, had the Kenosha page not been removed?

The next morning, as news outlets from across the United States reached out to Facebook for comment, he received his answer. The Kenosha page had violated Facebook's rules, but its moderators had not yet been trained in how to handle it.

Facebook's PR teams told journalists that Facebook had removed the Kenosha page and that the shooter was not a member of it. Both statements skirted the truth. While the shooter was not a member of the page, he was active on Facebook and had a history of browsing and commenting on pages dedicated to police forces. And while Facebook removed the event on the night of August 25, it had not actually taken action on the page for the Kenosha Guard. One of the members of the Kenosha Guard who had administrative privileges to the page removed it.

At the next day's Q&A, Zuckerberg said there had been an "an operational mistake" involving content moderators who hadn't been trained to refer these types of posts to specialized teams. He added that upon the second review, those responsible for dangerous organizations realized that the page violated their policies, and "we took it down." Zuckerberg's explanation didn't sit well with employees, who believed he was scapegoating content moderators, the lowest-paid contract workers on Facebook's totem pole. And when an engineer posted on a Tribe board that Zuckerberg wasn't being honest about who had actually removed the page, many felt they had been misled. "Once again, Mark was treating this like a one-off mistake. He was trying to blame others. He couldn't see that there was something systematically wrong with Facebook for a group like this to even exist, let alone create an event and have people posting stuff openly calling for people to be shot and killed," said one engineer, who repeated his concerns with his manager, only to be told to "be patient" while Facebook implemented its new policies.

But in the weeks that followed, right-wing groups began to rally around the Kenosha shooter. Facebook fan pages were created for him, and his actions were praised as those of a patriot. Facebook would remove more than 6,500 Facebook pages and groups dedicated to militias within the course of a month. As quickly as one page was taken down, several more would spring up in its place, circumventing the automated systems that were programmed to detect them.

The Kenosha crisis put a spotlight on all the platform's vulnerabilities. By pushing people into groups, Facebook had made it possible for fringe movements, including militias and conspiracy theorists, to organize and recruit followers on the site. Even when Facebook took the step of instituting an outright ban, many could slip through the cracks, with deadly consequences. Zuckerberg and Facebook would continue to deal with the groups on an individual basis, but that meant they were often taking action after a group had already become a problem.

A Tribe board comment from one of his longtime engineers was circulated to Zuckerberg through managers at the company. "If we break democracy, that is the only thing we will ever be remembered for," the engineer wrote in a post lamenting the vitriolic hate speech, partisan politics, and misinformation flooding Facebook. "Is that what we want our legacy to be?"

Over the course of six days in October 2020, Facebook made two big announcements. On October 6, it announced a comprehensive ban of QAnon from the site. And on October 12, it announced its policy against Holocaust misinformation. Quietly, in the background, it was removing thousands of militia groups. Zuckerberg was pivoting away from long-held beliefs on free speech, but no one at Facebook, including him, articulated it as a coherent change in policy. In a Q&A with employees on October 15, he pushed back on the idea that he was shifting his principles: "A bunch of people have asked, well, why are we doing this now? What is the reason and does this reflect a shift in our underlying philosophy? So I wanted to make sure that I got a chance to talk about this upfront. The basic answer is that this does not reflect a shift in our underlying philosophy or strong support of free expression. What it reflects is, in our view, an increased risk of violence and unrest."

Both Zuckerberg and Facebook's PR department framed the actions as individual decisions that happened to come at the same time. "Everyone at Facebook kind of accepted it, but that is just the kind of cult that is in Mark's inner circle. There was no one saying, 'Wait, have we changed our position on what we allow on Facebook? Should we explain what our thinking is?' No one stopped to do that," said one employee who participated in meetings with Zuckerberg at the time. "It was a month before the elections, and it felt like we were just in emergency mode."

Groups ranging from the ADL to the American Israel Public Affairs Committee praised the bans, but said they weren't enough; they pressed the platform to ban political ads the week ahead of the vote and to take strong action against misinformation, especially declarations of victory by any one candidate before official tallies were in. Zuckerberg agreed. Facebook had spent the last few years introducing features to target election misinformation, including working with outside fact-checkers, labeling content that was false or misleading, and taking down tens of thousands of Facebook accounts attempting to repeat the "coordinated inauthentic behavior" the Russians had attempted in 2016. It also discussed creating a "kill switch" for political advertising that would turn off political ads after November 3 if one candidate disputed the results. "It became very clear, the closer we got to the election, that we could have a situation on our hands where one candidate used Facebook to announce he was the victor," said a member of Facebook's election team, which spanned hundreds of people working across the company. "We didn't use the word *Trump*, but obviously we were talking about Trump." Zuckerberg had been participating in daily meetings with the election team, and he had told people close to him that November 3 was a "make-or-break" moment for the company. "He said that if we didn't get it right, no one would trust Facebook again," the election team member recalled.

No matter what steps the company took, Trump had amassed a Facebook following that was unyieldingly loyal and engaged. In the months leading up to the election, his audience on the platform had been swelling. In September alone, the president had nearly 87 million Facebook interactions on his page—more than CNN, ABC, NBC, the *New York Times*, the *Washington Post*, and Buzz-Feed combined.

Early on the morning of November 3, 2020, Alex Stamos paced around a T-shaped cluster of desks he had carefully arranged in the family room of his home. Two years after leaving Facebook, Stamos was still battling problems on the platform. At Stanford, where he had formed an Election Integrity Team that included academics, researchers, and other experts, he helped publish regular reports about foreign countries running disinformation networks on Facebook's platform and established a grading system for how Facebook, as well as other social media companies, was preparing for the elections. On the day of the vote, he was on high alert for anything, from hacked emails suddenly surfaced by a foreign state to an attack on election infrastructure in the United States. But the thing he was most concerned about was homegrown misinformation spread by Americans for Americans.

"It was obvious that our misinformation problem had become completely domestic," said Stamos, who noted that in the weeks and months leading up to the vote, an array of websites and media companies had been successfully seeding false narratives about the vote through social media. Over the course of the day, Stamos's team would file hundreds of reports to Facebook, Twitter, You-Tube, and other social media companies about misinformation it had found. Of all the companies, Facebook was the fastest to respond and remove problematic content.

Still, rumors swirled online that voting machines were changing people's ballots from Trump to Biden, that election officials were allowing people to vote under the names of their pet cats and dogs (or deceased relatives), and that certain states were purposefully miscounting votes. The rumors were baseless, but spread by President Trump and his supporters, they started to foster a narrative online that the election had, potentially, been stolen. "On Election Day and the day after, we saw a concerted effort by the president and his supporters in a way that was much more successful than what the Russians did in 2016," noted Stamos. "In a way, it was beyond what the Russians could have hoped for."

Facebook employees felt like they were fighting against an impossible tide of misinformation. They were being told to apply labels to any posts containing false information about the vote. But the labels, many of which simply directed people to Facebook's Voting Information Center, were ineffective. Few people read them or stopped sharing content that was flagged, the data showed. "Imagine if the tobacco industry put a label on cigarettes which said, 'Health officials have opinions about cigarettes. Go read this pamphlet about it!' That is the equivalent of what we were doing. We weren't coming out and saying, 'This is a lie. There is no evidence for this. Here is the truth,'" a member of Facebook's security team pointed out. "Or maybe, after years of allowing Trump to say whatever he wanted to say, people just got used to seeing crazy stuff on Facebook."

As the night wore on, and it became clear that the election would be too close to call, Facebook's election team realized its work was not done. Some of the remaining states could take days or even weeks to declare a winner. It was a scenario the team had prepared for, but had hoped would not happen. The day after the election, the team held a virtual meeting with Zuckerberg in which it requested he approve new break-glass emergency measures. These included an emergency change to the News Feed algorithm, which would put more emphasis on the "news ecosystem quality" scores, or NEQs. This was a secret internal ranking Facebook assigned to news publishers based on signals about the quality of their journalism. On one side of the spectrum were the outlets Facebook ranked as most trustworthy, such as the *New York Times* or the *Wall Street Journal*. On the other were outlets such as the Daily Caller, which had repeatedly been cited by Facebook's fact-checkers for publishing false or misleading stories. Under the change, Facebook would tweak its algorithms so people saw more content from the higher-scoring news outlets than from those that promoted false news or conspiracies about the election results.

Zuckerberg approved the change. For five days after the vote, Facebook felt like a calmer, less divisive space. "We started calling it 'the nicer News Feed,'" said one member of the election team. "It was this brief glimpse of what Facebook could be." Several people on the team asked whether the "nice News Feed" could be made permanent.

But by the end of the month, the old algorithm was slowly transitioned back into place. Guy Rosen, Facebook's head of integrity, confirmed that the changes were always meant to be temporary, an emergency measure taken in the midst of a critical election. Quietly, executives worried how conservatives would respond if a number of prominent right-wing outlets were permanently demoted. There was also concern that Facebook's changes had led to a decrease in sessions; users were spending less time on the platform. For the past year, the company's data scientists had been quietly running experiments that tested how Facebook users responded when shown content that fell into one of two categories: good for the world or bad for the world. The experiments, which were posted on Facebook under the subject line "P (Bad for the world)," had reduced the visibility of posts that people considered "bad for the world." But while they had successfully demoted them in the News Feed, therefore prompting users to see more posts that were "good for the world" when they logged into Facebook, the data scientists found that users opened Facebook far less after the changes were made.

The sessions metric remained the holy grail for Zuckerberg and the engineering teams. The group was asked to go back and tweak the experiment. They were less aggressive about demoting posts that were bad for the world, and slowly, the number of sessions stabilized. The team was told that Zuckerberg would approve a moderate form of the changes, but only after confirming that the new version did not lead to reduced user engagement.

"The bottom line was that we couldn't hurt our bottom line," observed a Facebook data scientist who worked on the changes. "Mark still wanted people using Facebook as much as possible, as often as possible."

Sheryl Sandberg sat in a sun-filled garden outside her Menlo Park home surrounded by high-resolution cameras set up to stream a live interview with the Reuters Next media conference. It was January 11, 2021, and within days, Joseph Biden would be sworn in as the forty-sixth president of the United States. But the country was still reeling from a week of turmoil in which hundreds of rioters had stormed the Capitol Building in Washington, DC, in what appeared to be an act of insurrection on behalf of Trump. With each passing day, reporters were uncovering more of the online footprint left behind by the rioters on social media and forming a picture of a violent network of right-wing extremists who believed they were serving the forty-fifth president.

Sandberg had been told that her interviewer would ask about Facebook's actions leading up to the riots. She had been prepped, by her advisers and comms staff, on how to answer. The company did not want a repeat of Zuckerberg's 2016 comments labeling Russian election interference "a pretty crazy idea." Highlight what Facebook has done to improve discourse on the platform and remove hate speech, Sandberg was told. Outline the huge efforts being made.

When asked about the events that had led to the siege on January 6, Sandberg suggested that the blame lay elsewhere, on new social media companies that had become home to the far right, such as Parler and Gab. "I think these events were largely organized on platforms that don't have our abilities to stop hate, don't have our standards, and don't have our transparency," she said. It was a quote that was picked up by news outlets across the world. Outraged members of Congress and researchers who studied right-wing groups accused Facebook of abdicating responsibility. "Sheryl Sandberg, Resign," read a headline in *The New Republic*.

Days later, indictments began to roll in for the rioters who had taken part in the attacks. As part of the legal filings, lawyers gathered evidence from Facebook, Twitter, and other social media accounts that laid bare how those organizing and participating in the Capitol storming had used Facebook's platform.

In one indictment, lawyers revealed how, in the weeks leading up to January 6, Thomas Caldwell and members of his militia group, the Oath Keepers, had openly discussed over Facebook the hotel rooms, airfare, and other logistics around their trip to Washington. On December 24, Caldwell had responded to a friend's

Facebook post by describing the route he was taking to DC. On December 31, he tried to rally his friends to join him by saying he was ready to "mobilize in the streets. This kettle is set to boil." On January 1, he discussed a possible hotel location in Washington that would allow him to "hunt" at night for members of the left-wing antifa group, whom he and other militias believed to be present in Washington.

Other indictments showed members of the Oath Keepers, Proud Boys, and other groups messaging one another on Facebook about their plans to carry weapons to Washington and the violent confrontations they hoped to have there. One Facebook page, called "Red-State Secession," posted on January 5, "If you are not prepared to use force to defend civilization, then be prepared to accept barbarism," to which a multitude of people posted comments, including pictures of weaponry they planned to bring to the president's rally on January 6.

Facebook's security and policy teams were aware of the activity and were growing increasingly alarmed. When journalists reported the Red-State Secession group to the company on the morning of January 6, an answer came back within hours that the security team had been reviewing the page and would remove it immediately. But while the platform was moving quickly to try to remove groups and pages calling for violence, they could not undo the simmering anger that for months had been building across thousands of Facebook pages. Under the banner of "Stop the Steal," a motto that underscored a wide-ranging allegation that the election had been stolen from President Trump, thousands of people had been mobilized to travel to Washington and take action.

Once in Washington, people freely celebrated on the morning of January 6 with posts on Facebook and Instagram showing the crowds that had gathered to hear President Trump deliver an address. Minutes after Trump ended his speech with a call to his supporters to "walk down Pennsylvania Avenue" toward the Capitol Building, where hundreds of members of Congress sat, people within the crowd used their phones to livestream clashes with police and the storming of the barricades outside the building. Many, including Caldwell, were getting messages on Facebook Messenger from allies watching their advance from afar.

"All members are in the tunnel under capital [sic]," read the message Caldwell received as he neared the building. Referring to members of Congress, the message added, "Seal them in. Turn on Gas."

Moments later, Caldwell posted a quick update on Facebook that read, "Inside."

He immediately began to get detailed instructions through an onslaught of Facebook messages that encouraged him to "take that bitch over." "Tom, all legislators are down in the Tunnels 3floors down," read one message. Another instructed Caldwell to go through each floor "from top to bottom" and gave him directions on which hallways to use. He was hunting for members of Congress, in a mission that he and other members of far-right groups across the United States saw as an act of insurrection and revolution.

Thousands of miles away, from their homes in the verdant suburbs surrounding MPK, Facebook executives watched with horror. On the advice of the security team, who warned that there was potential for violence in Washington that day, the group held a virtual meeting that morning to discuss contingency plans. It had been two months since the U.S. presidential election had ended, but executives felt like they had been holding their collective breath since November 3. "None of us had had a chance to exhale. We were still waking up every day feeling like the election wasn't yet over, and all eyes were on our response to Trump's unwillingness to concede the race to Biden," recalled one of the executives.

One of the biggest questions looming over the call was how Trump would use his Facebook account. Over the past week,

Facebook had begun to label many of his posts on voter fraud. But there was a question of what they would do if he pushed things further and declared via Facebook that he was not leaving office. The group also debated how they would respond if Trump directly called for the thousands gathered in Washington to engage in violence. Zuckerberg, while not present on the call, was briefed on their deliberations. At one point, the group floated getting Zuckerberg to call Trump to find out what the president would say. They ultimately decided against the move, out of concern that the conversation would likely leak to the press. It could make Facebook complicit in whatever Trump did that day.

Instead, the group had waited, watching to see whether Trump would repeat in a Facebook post the claim he was making in his speech—that the election had been stolen from him. They had watched as rioters stormed the Capitol and as breathless news anchors aired footage of members of Congress being rushed from the House floor by security. Facebook's security team informed them that some of the rioters were using their Facebook and Instagram accounts to livestream themselves wandering the Capitol halls. User reports of violent content had jumped more than tenfold since the morning, with almost forty thousand reports of violent content or comments coming in per hour. Zuckerberg encouraged the group to enact emergency measures and remove any of the violent videos showing rioters in the Capitol.

The group continued to watch as Trump cast blame on Vice President Pence for not going along with his, Trump's, scheme to strong-arm Congress into granting him a second term, and as he told the rioters he "loved" them. On Facebook, tens of thousands of people echoed the president's words, charging the rioters to hunt down the vice president.

As Zuckerberg and his executives debated what to do, a chorus was growing across different Tribe boards—engineers, product managers, designers, and members of the policy team all calling for Facebook to ban Trump from the platform, once and for all.

"It is no longer theoretical, he has actually incited violence," wrote one Facebook engineer. "I don't see how we justify keeping his account online."

"Hang in there everyone," Mike Schroepfer, Facebook's chief technology officer, wrote on a company-wide Tribe board.

"All due respect, but haven't we had enough time to figure out how to manage discourse without enabling violence?" responded one employee, one of many unhappy responses that swiftly garnered hundreds of likes from colleagues. "We've been fueling this fire for a long time and we shouldn't be surprised that it's now out of control."

By mid-afternoon Wednesday, as the last of the rioters was being escorted away from the Capitol, Zuckerberg had made up his mind to remove two of Trump's posts and to ban the president from posting for twenty-four hours.

It was the one step he had never wanted to take, and few in the group were clear on whether they were setting a precedent or responding to an unprecedented situation. The debate among his team had been exhaustive and intense. "At one point, someone told Zuckerberg that if we did this, we were going to have to answer the question of why now?" recalled one executive, who added that the group pointed out that in other countries, including the Philippines, India, and Ethiopia, Facebook had not removed or banned heads of government, even when their words appeared to incite violence. "So do we have one rule for America and another for everyone else? Are we saying we only act when it happens in America? It's not a good look."

Between the afternoon and nightfall, Zuckerberg waffled on what further steps he should take against Trump. By Thursday, he had decided that Facebook would extend its ban of Trump through the inauguration. He also ordered the security team to take sweeping action against a number of pro-Trump groups on the platform that had helped organize the January 6 rally, including the WalkAway campaign and hundreds of "Stop the Steal" groups.

"We believe that the public has a right to the broadest possible access to political speech," Zuckerberg posted on Facebook, explaining the decision. "But the current context is now fundamentally different, involving use of our platform to incite violent insurrection against a democratically elected government." He added that "the risks of allowing the President to continue to use our service during this period are simply too great."

It was the strongest action taken against Trump by any social media platform. But Zuckerberg ended the post with a mixed message: he was extending the ban on Trump's Facebook and Instagram accounts "indefinitely," which he went on to qualify as "at least the next two weeks until the peaceful transition of power is complete."

In other words, Facebook would keep its options open.

The Long Game

In May 2020, Facebook had announced the appointment of the first twenty members to an independent panel created to adjudicate the thorniest freedom of expression cases facing the platform. The Facebook Oversight Board was made up of global academics, comprising former political leaders and civil rights experts, who would choose from among public requests to appeal Facebook's decisions on content, including the removal of individual posts and user accounts. Zuckerberg had incubated the idea for years; his vision was that it would resemble a Supreme Court that would vote on cases and write out their opinions. The board was funded by Facebook but empowered to hand down binding rulings that couldn't be overturned by Zuckerberg or any of the company's leaders. "Facebook should not make so many important decisions about free expression and safety on our own," Zuckerberg explained in a post introducing the board.

One of the board's first decisions involved a post from Myanmar that Facebook initially removed because it was considered "pejorative or offensive" toward Muslims. In their ruling, the Oversight Board concluded that the post did not advocate hatred or incite imminent harm, though they acknowledged the dangerous role the platform had formerly played in the country.

The decision came at a strange time for Facebook, as the company had recently been cracking down on hate speech in Myanmar, expanding its staff of Burmese-speaking moderators, and working more closely with locals NGOs. In August 2020, nearly two years after the company deflected Matthew Smith's request to hand over potential evidence of war crimes by Myanmar's military to the UN, Facebook provided a cache of documents to the organization.

Many of the other cases the group selected early on were relatively minor complaints, ranging from the removal of a photo of a nipple to raise awareness for breast cancer to prescription cures for COVID-19 in France. But on January 21, 2021, Zuckerberg handed them a showstopper. Instead of making a final determination to ban Trump permanently after Biden's inauguration, he referred the case to the Facebook Oversight Board. It had until April to issue a ruling.

The Oversight Board offered Facebook the perfect out. Zuckerberg had effectively punted to others the tricky, and consequential, decision on what to do with Trump. It was a strategy of convenience that covered many fronts. In the wake of the Trump ban, civil society groups around the world demanded equal treatment of autocrats like Turkey's president Recep Tayyip Erdoğan and Venezuela's president Nicolás Maduro for inciting hate and spreading misinformation. How was the insurrection at the U.S. Capitol any worse than genocide and violence in Myanmar, Sri Lanka, or Ethiopia? Many employees and users also questioned why Trump was being punished for posts that incited violence in the Capitol riots, but not when he called for "shooting" in response to looters in Black Lives Matter protests. That was all in the board's court now.

Yet again, the company had figured out a way to abdicate responsibility, under the guise of doing what was best for the world.

It had no interest in enacting true reform, focusing instead on "performative changes," as Sen. Elizabeth Warren described the company's iterative decisions on political speech.

The federal and state antitrust lawsuits filed the previous December would be more difficult to dodge. The severity and breadth of the allegations and the call for Facebook's breakup had come as a surprise. Two months earlier, lawyers for the company had put out a paper arguing that if it were broken up, it would cost billions of dollars to maintain separate systems, would weaken security, and would spoil the experience of using the apps. "A 'breakup' of Facebook is thus a complete nonstarter," the lawyers concluded.

Over the summer, Zuckerberg and Sandberg had provided testimony to the FTC and states, and their aides thought their messages were airtight. Appearing in video testimony from her home, Sandberg casually kicked off her shoes and folded her legs under her, as she often does in meetings, and spooned the foam off her cappuccino while taking questions. She was loyal to the script: Facebook had lots of competitors and wasn't a monopoly. It had done everything possible to help Instagram and WhatsApp thrive within the company. As had been the case with FTC officials a decade earlier, Sandberg's casualness took some regulators by surprise and showed her misreading the seriousness of their intent. It was as if she were having a chat with friends, one person recalled. For all the talk of her vaunted political instincts, time and again, she revealed herself to be curiously oblivious and overconfident.

Facebook itself wasn't taking any chances, however. The company had hired former FTC officials to lobby on its behalf to antitrust officials and to write white papers supporting its defense. The company had more than one hundred lawyers in-house and on retainer, including the lead trial attorneys from white-shoe litigation firm Kellogg, Hansen and a former FTC general counsel, now at Sidley Austin. Facebook's in-house attorneys believed the FTC and the states had a flimsy case that rested on a novel theory of antitrust—that Facebook wanted to squash a future competitor—and the company was willing to bring the full weight of its power and resources to the court battle. The FTC was essentially calling for a do-over of the mergers, which the commission hadn't opposed when it vetted the deals years ago. "So, they are basically saying to a court they were wrong and 'listen to us now'?" one Facebook official who worked on the antitrust team said.

Zuckerberg was careful not to express his opinions. In a letter to employees, he cautioned that they should not discuss legal matters publicly. But he signaled his willingness to fight for the long term. The lawsuits were just "one step in a process which could take years to play out." At the same time the suits landed, Facebook resorted to its usual tactics of deflection and defensiveness, pointing to bigger threats at other companies. Its lawyers prepared a possible antitrust suit against Apple for new privacy restrictions on apps like Facebook. Executives continued to warn of the threat of China and a potential law in Australia that would require Facebook to pay news publishers to host articles on the social network.

In the meantime, the DC office hustled to accommodate the changing of the guard in Washington. At the start of every new administration, the Capitol's lobbying sub-economy reboots, with companies reorienting their personnel and priorities to cater to the new party in charge. For Facebook, the change was like the ocean-churning 180-degree turn of a cruise ship. The company had tilted so far to please Trump that Zuckerberg's effusive embrace of the Biden administration struck a dissonant chord for employees and political leaders. And it required a significant overhaul of its colossal lobbying operation.

But Facebook's Washington office was hardly an underdog. The company had outspent nearly every other company to build up its defenses. In 2020, it ranked second among individual corporations across all sectors in lobbying, shelling out nearly twenty million dollars and outspending Amazon, Alphabet, Big Oil, Big Pharma, and retail. The years of nurturing relationships in both parties and ties to prominent business and political leaders paid their own dividends. Former board member Jeff Zients was appointed to lead Biden's COVID task force. Former board member Erskine Bowles played a role in Biden's transition team, as did Jessica Hertz, a Facebook attorney who had worked on the Cambridge Analytica investigation and who later was named the president's staff secretary.

Democrats within the DC office jockeyed for the top positions of directly lobbying the White House and leading the team of congressional lobbyists. Though Clegg knew Biden from his time as the United Kingdom's deputy prime minister, Sandberg was again the most powerful Democrat in the company. During the transition, a few press reports speculated that she could leave for a position in the Biden administration, possibly as a cabinet member. But her reputation and Facebook's brand were too toxic, some Democratic operatives believed. Or, as one Biden transition adviser put it, "No fucking way."

For some, decisions made under Joel Kaplan during the Trump administration had finally worn them out. Democratic lobbyist Catlin O'Neill left in January, telling friends her decision was based on an accumulation of actions she disagreed with, including the decision not to remove the manipulated video of her old boss, Speaker Pelosi. Others at the company felt more comfortable criticizing Kaplan and his track record. "If Joel is so good, why is our CEO testifying again?" an employee who worked on policy issues quipped when Zuckerberg was called before the Senate Judiciary Committee for a hearing on social media censorship days before the election. "Any other company would have fired their whole Washington office."

But Kaplan's job was secure. Even with a Democratic White

House and Democratic majority in Congress, he continued to serve as vice president of global public policy. He had made it inside Zuckerberg's inner circle. Kaplan would lie low in the new administration, and others would have more direct contact with the Biden White House, but he continued to have Zuckerberg's ear on all political matters. "Joel will go when Joel is ready to go. Mark trusts very few people on policy, and Joel is in the center of that circle of trust," the policy team member said. Cox and Boz remained firmly ensconced in that circle as well.

Also secure was Sandberg and Zuckerberg's relationship. She had fallen short of his expectations at times, but Clegg absorbed some of the heat as he embraced the role of ambassador to one of the most chastised companies in the world—a role Sandberg was happy to shed, according to confidantes. She wasn't interested in a shift into politics, people close to her said; she felt there was still too much she needed to fix at Facebook. She was also content with her personal life and not interested in any kind of disruption for her middle school–age children; she and her fiancé, Tom Bernthal, had combined their families. Berthnal's three children had moved from Los Angeles to Menlo Park and enrolled in school there. Some employees claimed that they weren't seeing her as often in high-level meetings, but Sandberg's aides insisted this was due to a change in meeting schedules because of leaks to the press and working-from-home conditions due to COVID-19. Gossip aside, there was no question that in one significant aspect, she continued to hold up her end of the partnership: the constant churn of profits from the ad business she had masterminded.

On January 27, 2021, Zuckerberg and Sandberg displayed their unique dynamic in an earnings call with investment analysts, during which they delivered two very different messages. In yet another about-face decision on speech, Zuckerberg announced that Facebook was planning to deemphasize political content in the News

Feed because, he said, "people don't want politics and fighting to take over their experience on our service." He was still making calls on the biggest policy decisions. The announcement was also a tacit acknowledgment of Facebook's years-long failure to control hazardous rhetoric running roughshod on the social network, particularly during the election. "We're going to focus on helping millions of more people participate in healthy communities, and we're going to focus even more on being a force for bringing people closer together," he added.

Then Sandberg shifted the focus to earnings. "This was a strong quarter for our business," she said. Revenue for the fourth quarter was up 33 percent, to $28 billion, "the fastest growth rate in over two years." During the pandemic, users were on the site more than ever—2.6 billion people used one of Facebook's three apps every day—and advertisers were clamoring to reach them.

By the time you read this book, Facebook could look very different. Zuckerberg may step away from the job of CEO to spend more time on his philanthropic endeavors. The way people connect on Facebook may not be on a phone but on some other device, like augmented reality headsets. The company's most popular utility may not be status updates and shares, but something like blockchain payments for retail goods or the production and distribution of blockbuster entertainment.

With $55 billion in cash reserves, the company has endless options to buy or innovate its way into new lines of business, as Google has with autonomous vehicles and Apple with health devices. Even during his annus horribilis of 2020, Zuckerberg was looking toward the future. In a quest to break into the lucrative field of corporate communications software, in late November, Facebook bought Kustomer for $1 billion. The popularity of the Zoom tool during the pandemic had rankled, and Zuckerberg challenged employees to come up with a videoconferencing rival. He directed more engineering resources to expand functions on virtual reality and augmented reality headsets, which he described as the new frontier for how people would communicate in the future. The company was also toying with publishing tools for users, fifteen years after the *Washington Post*'s Donald Graham offered to buy a stake in the tech start-up. And despite pressure from regulators, Zuckerberg was committed to developing its Libra blockchain currency project, which it rebranded as "Diem."

Throughout Facebook's seventeen-year history, the social network's massive gains have repeatedly come at the expense of consumer privacy and safety and the integrity of democratic systems. And yet, that's never gotten in the way of its success. Zuckerberg and Sandberg built a business that has become an unstoppable profit-making machine that could prove too powerful to break up. Even if regulators, or Zuckerberg himself, decided to one day end the Facebook experiment, the technology they have unleashed upon us is here to stay.

One thing is certain. Even if the company undergoes a radical transformation in the coming years, that change is unlikely to come from within. The algorithm that serves as Facebook's beating heart is too powerful and too lucrative. And the platform is built upon a fundamental, possibly irreconcilable dichotomy: its purported mission to advance society by connecting people while also profiting off them. It is Facebook's dilemma and its ugly truth.

Acknowledgments

This book would not be possible without the many sources who trusted us to tell their stories. They spoke to us often at great personal and professional risk. We are so grateful for their participation, patience, and commitment to the pursuit of truth. While some have since left Facebook, many remain at the company and are trying to change things from within.

Thank you to the entire team at Harper for believing in this book and for taking on the enormous task of shepherding a story that was unfolding in real time and written by two authors on opposite coasts. Jennifer Barth, our editor, brought commitment to the subject and rallied the whole publishing house, led by Jonathan Burnham, behind a project that took several twists and turns as the Facebook story evolved. We appreciate Jennifer's dedication, curiosity, ideas, and tenacity with a very hard project.

The team at Little, Brown, led by Tim Whiting, were unflagging in their enthusiasm for this book and our work. They brought an international perspective and reminded us to keep an eye on the repercussions of Facebook's global reach.

We are also grateful to the *New York Times*, where this book germinated as a series of stories with many colleagues across desks and with the full support of newsroom leaders. Firstly, we thank our dogged and supportive editor, Pui-Wing Tam, who brought a simple question to us in the summer of 2017: What is going on inside Facebook as it slides into one scandal after another? She was not satisfied with a cursory look at the company. She pushed us to get more and to go deeper. Her sharp questions and incisive thinking have made us better journalists.

Our agents, Adam Eaglin and Elyse Cheney, stood with us every step of the way. They believed there needed to be a book about Facebook and for the world to understand this powerful company and force in society. They read chapters and outlines, and went far beyond our expectations to make this—the first book for both of us—the best manuscript possible. We are also grateful to the entire crew at the Cheney Agency for promoting our work across the globe. Thank you to Isabel Mendia, Claire Gillespie, Alice Whitwham, Allison Devereux, and Danny Hertz.

Rebecca Corbett not only guided us in the *New York Times* newsroom, but was an important voice in the book process with her famously big ideas, news judgment, skepticism, and deft hand at structure, themes, and characters. After full days working on blockbuster newspaper investigations, Rebecca started all over again in the evenings and on weekends to talk to us about chapters and ideas. Gareth Cook came earlier in the process, helping us sift through volumes of material to see the bigger picture and plot out our plan. His passion for the topic and deft hand at structure and arguments brought our themes into sharper view.

Our fact checkers, Hilary McClellan and Kitty Bennett, who also did research for us, were invaluable. Dealing with many hundreds of pieces of information, they were meticulous and committed to the highest standards.

Beowulf Sheehan performed a bit of magic with our author photo. We were unable to meet in person but Beowulf took our photos remotely and then melded them together with flawless execution. We thank Beowulf for his incredible artistry, patience, and sense of adventure.

We were buffeted by the incredible reporting of our *New York Times* colleagues, too many to name in full. But in summary, Nicholas Confessore, Matthew Rosenberg, and Jack Nicas were our "OG" crew, working together on a story in November 2018 that struck the public conscience in a way we have rarely experienced. Mike Isaac, Kevin Roose, Scott Shane, and others were part of a formidable reporting crew dedicated to understanding the power of Facebook. We are indebted to the rest of the technology pod, a team of incredibly talented and hard-working reporters dedicated to holding Silicon Valley's giants to account. We leaned heavily on their reporting and we are grateful for their collegiality. Our editors, Joe Plambeck and James Kerstetter, were champions of our reporting at the paper and on this project. *New York Times* leadership—Dean Baquet, Joe Kahn, Matt Purdy, Rebecca Blumenstein and Ellen Pollock—offered important support and made this book possible by allowing us to take leaves to report and write. A. G. Sulzberger's kind notes on our coverage were always a welcome surprise in our inbox, and his enthusiasm for our tech coverage has been inspiring.

This book also builds on the reporting of many other journalists who have tirelessly worked to shed light on the company. To name a few: Ryan Mac, Craig Silverman, Sarah Frier, Deepa Seetharaman, Casey Newton, Julia Angwin, Kara Swisher, David Kirkpatrick, Steven Levy, Jeff Horowitz, Lizza Dwoskin, Julia Carrie Wong, Brandy Zadrozny, and Ben Collins.

Our readers—Kashmir Hill, Jessica Garrison, Kevin Roose, Natasha Singer, Scott Shane, and Brian Chen—were incredibly generous. They offered pages of feedback—from the most abstract and philosophical to specific challenges to reporting and our ideas. All of their feedback has been incorporated one way or another into the final version of the book.

From Cecilia: This book starts and ends with the support of my loving family. They endured far too many nights and weekends without me as I holed up to write. Oltac, my dear, is my champion. He rallied behind my ideas and helped shape my understanding of markets and business and political philosophy. My darling children, Leyla and Tigin, were my sharpest editors with their unvarnished feedback on narrative and style. They cheered me on and had so much faith in me. My parents, William and Margaret, are my true North. Dad has binders of every article I've written, starting with clips from Seoul when I was a cub reporter. Their journeys remind me that writing is a great privilege; they made my career possible. Sabrina, Felicia, Grace and Drew scrutinized chapters, got emotionally invested in the subject and fueled my ambitions. Felicia brought intellectual vigor to discussions on themes related to American history and the Classics. They reminded me that telling the story of Facebook was important. This book is for my family.

From Sheera: To my family, who always believed I could write a book: We did it! I would not be a journalist, or even a writer, if my first champion, Grandpa Louis, had not instilled in me a love of newspapers. I wrote this book at his old roll top desk. My grandmothers, Yona and Malka, would have cracked an egg on my head for luck. My parents, Mike and Irit, raised me to challenge, ask questions, and never back down. Talia, who brings creativity and passion to every discussion, was as excited about the book as I was. Elan, who always thinks outside the box, encouraged me to always consider different perspectives.

Tom, my love, makes all my dreams possible. He gave me the space, time, and support I needed to write this book. I started reporting the *New York Times* article that became this book when I was pregnant with Ella in 2017, and worked on the opening chapters when pregnant with Eden. I am happy that neither of their first words was "Facebook." I may have carried Ella and Eden through this process, but they sustained me. Mama loves you ad hashamayim.

Notes

Prologue: At Any Cost

1   New York State Attorney General Letitia James: "NY Attorney
    General Press Conference Transcript: Antitrust Lawsuit against Facebook,"
    December 9, 2020.

1   amounted to a sweeping indictment: *State of New York et al. v.
    Facebook, Inc.*, antitrust case filed in the United States District Court for
    the District of Columbia, Case 1:20-cv-03589-JEB, Document 4, filed
    December 9, 2020. https://ag.ny.gov/sites/default/files/state_of_new
    _york_et_al._v._facebook_inc._-_filed_public_complaint_12.11.2020.pdf.

3   the words of academic and activist Shoshana Zuboff: John Naughton,
    "'The Goal is to Automate Us': Welcome to the Age of Surveillance
    Capitalism," *Observer*, January 20, 2019.

4   Zuckerberg and Sandberg met at a Christmas party: Elise Ackerman,
    "Facebook Fills No. 2 Post with Former Google Exec," *Mercury News*,
    March 5, 2008.

4   $85.9 billion in revenue in 2020: Facebook, "Facebook Reports Fourth
    Quarter and Full Year 2020 Results," press release, January 27, 2021.

Chapter 1: Don't Poke the Bear

7   with ads recently expanded on Instagram: Vindu Goel and Sidney
    Ember, "Instagram to Open Its Photo Feed to Ads," *New York Times*,
    June 2, 2015.

8   spying on user data as it sat unprotected: Barton Gellman and Ashkan
    Soltani, "Russian Government Hackers Penetrated DNC, Stole Opposition
    Research on Trump," *Washington Post*, October 30, 2013.

11 Trump promised to take: Jenna Johnson, "Donald Trump Calls for
    'Total and Complete Shutdown of Muslims Entering the United States,'"
    Washington Post, December 7, 2015,

12 "Don't poke the bear": Sheera Frenkel, Nicholas Confessore, Cecilia Kang, Matthew Rosenberg, and Jack Nicas, "Delay, Deny and Deflect: How Facebook's Leaders Fought through Crisis," *New York Times*, November 14, 2018.

15 From the start of Trump's presidential campaign: Issie Lapowsky, "Here's How Facebook *Actually* Won Trump the Presidency," *Wired*, November 2016.

15 Facebook employees who were embedded: Sarah Frier and Bill Allison, "Facebook 'Embeds' Helped Trump Win, Digital Director Says," *Bloomberg*, October 6, 2017.

15 bought thousands of postcard-like ads and video messages: Andrew Marantz, "The Man Behind Trump's Facebook Juggernaut," *New Yorker*, March 9, 2020.

15 By early 2016, 44 percent: Jeffrey Gottfried, Michael Barthel, Elisa Shearer and Amy Mitchell, "The 2016 Presidential Campaign—a News Event That's Hard to Miss," Journalism.org, February 4, 2016.

16 An employee stepped up to a microphone stand: Deepa Seetharaman, "Facebook Employees Pushed to Remove Trump's Posts as Hate Speech," *Wall Street Journal*, October 21, 2016.

17 "being co-opted and twisted": Renée DiResta, "Free Speech Is Not the Same as Free Reach," *Wired*, August 30, 2018.

Chapter 2: The Next Big Thing

21 two student groups at Harvard: Katharine A. Kaplan, "Facemash Creator Survives Ad Board," *Harvard Crimson*, November 19, 2003.

22 the *Harvard Crimson* slammed the project: Laura L. Krug, "Student Site Stirs Controversy," *Harvard Crimson*, March 8, 2003.

22 combining his in-development social network with Greenspan's project: Aaron Greenspan, "The Lost Chapter," AaronGreenspan.com, September 19, 2012.

23 "I kind of want to be the new MTV": Claire Hoffman, "The Battle for Facebook," *Rolling Stone*, September 15, 2010.

24 In one online chat: Nicholas Carlson, "'Embarrassing and Damaging' Zuckerberg IMs Confirmed by Zuckerberg, *New Yorker*," *Business Insider*, September 13, 2010.

25 Just six months earlier, he had moved: Erica Fink, "Inside the 'Social Network' House," CNN website, August 28, 2012.

25 ideology was rooted in a version of libertarianism: Noam Cohen, "The Libertarian Logic of Peter Thiel," *Wired*, December 27, 2017.

25 In 2011, Thiel would endow a fellowship: MG Siegler, "Peter Thiel Has New Initiative to Pay Kids to 'Stop out of School,'" TechCrunch, September 27, 2010.

27   The site had a few ads from local Cambridge businesses: Seth Fiegerman, "This is What Facebook's First Ads Looked Like," *Mashable*, August 15, 2013.

29   By the end of 2004, one million college students: Anne Sraders, "History of Facebook: Facts and What's Happening," TheStreet, October 11, 2018.

30   The board at the time: Allison Fass, "Peter Thiel Talks about the Day Mark Zuckerberg Turned down Yahoo's $1 Billion," *Inc.*, March 12, 2013.

30   The investment and buyout offers kept coming in: Nicholas Carlson, "11 Companies that Tried to Buy Facebook Back When it Was a Startup," *Business Insider*, May 13, 2010.

31   The Yahoo buyout offer: Fass, "Peter Thiel Talks about the Day Mark Zuckerberg Turned down Yahoo's $1 Billion."

31   "It was the first point where we had to look at the future": Mark Zuckerberg's August 16, 2016 interview with Sam Altman, "How to Build the Future," can be viewed on YouTube.

32   he spent most of his time working on an idea: Stephen Levy, "Inside Mark Zuckerberg's Lost Notebook," *Wired*, February 12, 2020.

32   Zuckerberg imagined a personalized hierarchy of "interesting-ness": Steven Levy, *Facebook: The Inside Story* (New York: Blue Rider Press, 2020).

34   But the morning brought angry users: Tracy Samantha Schmidt, "Inside the Backlash against Facebook," *Time*, September 6, 2006.

35   "When we watched people use it": UCTV's "Mapping the Future of Networks with Facebook's Chris Cox: The Atlantic Meets the Pacific," October 8, 2012, can be viewed on YouTube.

37   Twitter, which had launched in July 2006: Taylor Casti, "Everything You Need to Know about Twitter," *Mashable*, September 20, 2013.

38   he reflected on his early experiences as a CEO: Zuckerberg, quoted in *Now Entering: A Millennial Generation*, directed by Ray Hafner and Derek Franzese, 2008.

Chapter 3: What Business Are We In?

41   she married Brian Kraff: Marital Settlement Agreement between Sheryl K. Sandberg and Brian D. Kraff, Florida Circuit Court in Dade County, filed August 25, 1995.

42   Sandberg had traveled with Summers: Sandberg, interview with Reid Hoffman, *Masters of Scale* podcast, October 6, 2017,

42   "Sheryl, don't be an idiot": Ibid.

43   "I lost the coin flip as to where we were going to live": Peter Holley, "Dave Goldberg, Husband of Facebook Exec Sheryl Sandberg, Dies Overnight, Family Says," *Washington Post*, May 2, 2015.

43   Sandberg thrived: Brad Stone and Miguel Helft, "Facebook Hires a Google Executive as No. 2," *New York Times*, March 5, 2008.

43   She met Zuckerberg at a Christmas party: Patricia Sellers, "The New Valley Girls," *Fortune*, October 13, 2008.

44   "That they would cross paths was inevitable": Kara Swisher, "(Almost) New Facebook COO Sheryl Sandberg Speaks!" AllThingsD, March 10, 2008.

46   Federal Trade Commission issued self-regulatory principles: FTC, "FTC Staff Proposes Online Behavioral Advertising Principles," press release, December 20, 2007.

47   Zuckerberg accompanied Sandberg: Henry Blodget, "The Maturation of the Billionaire Boy-Man," *New York* magazine, May 4, 2012.

47   "Mr. Zuckerberg and Ms. Sandberg will face mounting pressure": Vauhini Vara, "Facebook CEO Seeks Help as Site Grows," *Wall Street Journal*, March 5, 2018.

49   in a winding Facebook post: Andrew Bosworth, Facebook post, December 1, 2007.

49   demeaning comments casually made about women: Katherine Losse, *The Boy Kings* (New York: Simon and Schuster, 2012), p. 24.

49   On Sandberg's first day: Bianca Bosker, "Mark Zuckerberg Introduced Sheryl Sandberg to Facebook Staff by Saying They Should All 'Have a Crush on' Her," Huffington Post, June 26, 2012, https://www.huffpost.com/entry/mark-zuckerberg-sheryl-sandberg-facebook-staff-crush_n_1627641.

51   A month into her new job: David Kirkpatrick, *The Facebook Effect* (New York: Simon and Schuster, 2010), p. 257, and interviews.

52   In 1994, an engineer at Netscape had created the "cookie": Janet Guyon, "The Cookie that Ate the World," Techonomy, December 3, 2018.

53   ad business was essentially outsourced to Microsoft: Katharine Q. Seelye, "Microsoft to Provide and Sell Ads on Facebook, the Web Site," *New York Times*, August 23, 2006.

53   Sandberg had been in charge: Francesca Donner, "The World's Most Powerful Women," *Forbes*, August 19, 2009.

55   "Sheryl people" or "Mark people": Josh Constine, "How the Cult of Zuck Will Survive Sheryl's IPO," TechCrunch, March 1, 2012.

56   she began wooing the biggest brands: Jessica Guynn, "Facebook's Sheryl Sandberg Has a Talent for Making Friends," *Los Angeles Times*, April 1, 2012.

56   "Unlike any other executive": Ibid.

57   the company's latest innovation: Louise Story and Brad Stone, "Facebook Retreats on Online Tracking," *New York Times*, November 30, 2007.

59   MoveOn.org circulated a petition: Ibid.

59   Coca-Cola and Overstock dropped out: John Paczkowski, "Epicurious Has Added a Potential Privacy Violation to Your Facebook Profile!," AllThingsD, December 3, 2007.

60   In a blog post, he assured: Mark Zuckerberg, "Announcement: Facebook Users Can Now Opt-Out of Beacon Feature," Facebook blog post, December 6, 2007.

60   entered a new phase: Donner, "The World's Most Powerful Women."

61   "What we believe we've done": Sandberg, "Welcome to the Cloud" panel, Dreamforce 2008 conference, San Francisco, FD (Fair Disclosure) Wire, November 3, 2008.

61   Facebook's success "depends upon one-way-mirror operations": Shoshana Zuboff, "You Are Now Remotely Controlled," New York Times, January 24, 2020.

62   "If you have something that you don't want anyone to know about": Julie Bort, "Eric Schmidt's Privacy Policy is One Scary Philosophy," Network World, December 11, 2009.

64   Twitter was a loud, noisy and increasingly crowded town square: Leena Rao, "Twitter Added 30 Million Users in the Past Two Months," TechCrunch, October 31, 2010.

64   In December 2009, he announced a gutsy move: Bobbie Johnson, "Facebook Privacy Change Angers Campaigners," Guardian, December 10, 2009.

64   "In short, this is Facebook's answer to Twitter": Jason Kincaid, "The Facebook Privacy Fiasco Begins," TechCrunch, December 9, 2009.

65   Sparapani defended the action: Cecilia Kang, "Update: Questions about Facebook Default for New Privacy Rules," Washington Post, December 9, 2009.

65   Facebook's definition of privacy: Ryan Singel, "Facebook Privacy Changes Break the Law, Privacy Groups Tell FTC," Wired, December 17, 2009.

65   sharing online was becoming a "social norm": Bobbie Johnson, "Privacy No Longer a Social Norm, Says Facebook Founder, Guardian, January 11, 2010.

67   Center for Digital Democracy and nine other privacy groups: In the Matter of Facebook, Inc., EPIC complaint filed with the FTC on December 17, 2009, https://epic.org/privacy/inrefacebook/EPIC-FacebookComplaint.pdf.

Chapter 4: The Rat Catcher

69   "Mark Zuckerberg Asks Racist Facebook Employees": Michael Nuñez, "Mark Zuckerberg Asks Racist Facebook Employees to Stop Crossing out Black Lives Matter Slogans," Gizmodo, February 25, 2016.

71   It was a memo to all employees: Ibid.

71   The story, published on February 25, at 12:42 p.m.: Ibid.

78   And the article published as a result: Michael Nuñez, "Former Facebook Workers: We Routinely Suppressed Conservative News," Gizmodo, May 9, 2016.

78   "It is beyond disturbing to learn": John Herrman and Mike Isaac, "Conservatives Accuse Facebook of Political Bias," *New York Times*, May 9, 2016.

79   rumored to be on a short list for a cabinet position: Zachary Warmbrodt, Ben White, and Tony Romm, "Liberals Wary as Facebook's Sandberg Eyed for Treasury," Politico, October 23, 2016.

81   On May 18, sixteen prominent conservative: Mike Isaac and Nick Corasaniti, "For Facebook and Conservatives, a Collegial Meeting in Silicon Valley," *New York Times*, May 18, 2016.

82   Beck later wrote on his blog: Brianna Gurciullo, "Glen Beck on Facebook Meeting: 'It Was Like Affirmative Action for Conservatives,'" Politico, May 19, 2016.

82   Beck had falsely accused a Saudi national: Daniel Arkin, "Boston Marathon Bombing Victim Sues Glenn Beck for Defamation," NBC News website, April 1, 2014.

84   "We connect people": Ryan Mac, Charlie Warzel and Alex Kantrowitz, "Growth at Any Cost: Top Facebook Executive Defended Data Collection in 2016 Memo—and Warned that Facebook Could Get People Killed," Buzzfeed News, March 29, 2018.

86   When, in June 2012: Facebook, "Facebook Names Sheryl Sandberg to Its Board of Directors," press release, June 25, 2012.

87   "Her name has become a job title": Miguel Helft, "Sheryl Sandberg: The Real Story," *Fortune*, October 10, 2013.

87   She was cultivating new ad tools: Keith Collins and Larry Buchanan, "How Facebook Lets Brands and Politicians Target You," *New York Times*, April 11, 2018.

87   It was also working on a tool called "Lookalike Audiences": David Cohen, "Facebook Officially Launches Lookalike Audiences," *Adweek*, March 19, 2013.

88   "Reading the questions was a painful but productive exercise": "Facebook Executive Answers Reader Questions," *New York Times* "Bits" blog, May 11, 2010.

89   Since opening the platform: Sarah Perez, "More Cyberbullying on Facebook, Social Sites than Rest of the Web," *New York Times*, May 10, 2010.

89   *Consumer Reports* estimated that 7.5 million: "CR Survey: 7.5 Million

Facebook Users are Under the Age of 13, Violating the Site's Terms," *Consumer Reports*, May, 2011.

92   As early as 2008, the company had been battling: Lisa Belkin, "Censoring Breastfeeding on Facebook," *New York Times* "Motherlode" blog, December 19, 2008.

Chapter 5: The Warrant Canary

98   On June 14, CrowdStrike: Editorial Team, "CrowdStrike's Work with the Democratic National Committee: Setting the Record Straight," CrowdStrike, June 5, 2020.

98   ThreatConnect and Secureworks: ThreatConnect Research Team, "Does a BEAR Leak in the Woods?" ThreatConnect, August 12, 2016. Secureworks Counter Threat Unit, "Threat Group 4127 Targets Hillary Clinton Presidential Campaign," Secureworks, June 16, 2016.

99   Russia had been caught doping: Motez Bishara, "Russian Doping: 'An Unprecedented Attack on the Integrity of Sport & the Olympic Games,'" CNN website, July 18, 2016.

100   Wasserman Schultz was forced to resign: Jonathan Martin and Alan Rappeport, "Debbie Wasserman Schultz to Resign D.N.C. Post," *New York Times*, July 24, 2016.

100   The Podesta emails, which highlighted mudslinging: Scott Detrow, "What's in the Latest WikiLeaks Dump of Clinton Campaign Emails," NPR, October 12, 2016.

101   Stamos was Yahoo's information security officer: Arik Hesseldahl, "Yahoo to Name TrustyCon Founder Alex Stamos as Next Chief Information Security Officer," Vox, February 28, 2014.

102   he discovered that the vulnerability: Joseph Menn, "Yahoo Scanned Customer Emails for U.S. Intelligence," Reuters, October 4, 2016.

105   "Russia, if you're listening": Michael S. Schmidt, "Trump Invited the Russians to Hack Clinton. Were They Listening?," *New York Times*, July 13, 2018.

106   his campaign was busily buying up millions: Ian Bogost and Alexis C. Madrigal, "How Facebook Works for Trump," *Atlantic*, April 17, 2020.

106   Over the last few years, the campaigns: Davey Alba, "How Duterte Used Facebook to Fuel the Philippine Drug War," Buzzfeed News, September 4, 2018.

108   The Fancy Bear hackers had stolen: Ben Chapman, "George Soros Documents Published 'by Russian Hackers' say US Security Services," *Independent*, August 15, 2016.

108   in November 2015, Russia's Office of the Prosecutor-General: Jennifer Ablan, "Russia Bans George Soros Foundation as State Security 'Threat'," Reuters, November 30, 2015.

Chapter 6: A Pretty Crazy Idea

112 Among them were Trump's former: Robert Costa, "Former Carson Campaign Manager Barry Bennett is Quietly Advising Trump's Top Aides," *Washington Post*, January 22, 2016.

113 He had been accused of misdemeanor battery: Kerry Saunders and Jon Schuppe, "Authorities Drop Battery Charges against Trump Campaign Manager Corey Lewandowski," NBC News website, April 14, 2016.

115 "I think the idea that fake news on Facebook": Adrienne Jane Burke, "Facebook Influenced Election? Crazy Idea, Says Zuckerberg," Techonomy, November 11, 2016.

Chapter 7: Company over Country

121 Trump received briefings: Salvador Rizzo, "Did the FBI Warn the Trump Campaign about Russia?" *Washington Post*, September 20, 2019.

124 His earliest speechwriter, Kate Losse, wrote: Kate Losse, "I Was Zuckerberg's Speechwriter. 'Companies over Countries' Was His Early Motto," Vox, April 11, 2018.

142 "The American people deserve to see": Kyle Cheney and Elana Schor, "Schiff Seeks to Make Russia-linked Facebook Ads Public," Politico, October 2, 2017.

Chapter 8: Delete Facebook

149 "The breach allowed the company": Matthew Rosenberg, Nicholas Confessore and Carole Cadwalladr, "How Trump Consultants Exploited the Facebook Data of Millions," *New York Times*, March 17, 2018.

149 "the unprecedented data harvesting": Carole Cadwalladr and Emma Graham-Harrison, "Revealed: 50 Million Facebook Profiles Harvested for Cambridge Analytica in Major Data Breach," *Guardian*, March 17, 2018.

150 up to 87 million Facebook users: Rosenberg, Confessore and Cadwalladr, "How Trump Consultants Exploited the Facebook Data of Millions."

150 Three weeks later, Zuckerberg sat: "Facebook CEO Mark Zuckerberg Testifies on User Data," April 10, 2018, Video can be found on C-Span .org.

151 How Cambridge Analytica had breached Facebook users' privacy: *State of New York et al. v. Facebook.*

152 In 2012, Parakilas alerted senior executives: Paul Lewis, "'Utterly Horrifying': Ex-Facebook Insider Says Covert Data Harvesting Was Routine," *Guardian*, March 20, 2018.

152 In 2014, Zuckerberg shifted strategies: CPO Team, "Inside the Facebook Cambridge Analytica Data Scandal, CPO Magazine, April 22, 2018.

152 multiplying his data set to nearly 90 million Facebook users: Cecilia
Kang and Sheera Frenkel, "Facebook Says Cambridge Analytica Harvested
Data of Up to 87 Million Users," *New York Times*, April 4, 2018.

153 Klobuchar of Minnesota called for Zuckerberg to testify: Brooke Seipel and
Ali Breland, "Senate Judiciary Dem Calls on Zuckerberg to Testify before
Committee," The Hill, March 17, 2018.

153 Kennedy of Louisiana joined Klobuchar: Reuters staff, "Republican
Senator Joins Call for Facebook CEO to Testify about Data Use," Reuters,
March 19, 2018.

153 Cher announcing the next day: Cher, "2day I did something very hard
4me," Tweet posted March 20, 2018.

153 On Tuesday, March 20, 2018: Casey Newton, "Facebook Will hold
an Emergency Meeting to Let Employees Ask Questions about Cambridge
Analytica," The Verge, March 20, 2018.

153 That same day, members of the Federal Trade Commission: Cecilia
Kang, "Facebook Faces Growing Pressure over Data and Privacy Inquiries,"
New York Times, March 20, 2018.

153 that settlement had resolved a complaint: FTC, "Facebook Settles FTC
Charges that it Deceived Consumers by Failing to Keep Privacy Promises,"
press release, November 29, 2011.

154 British authorities also opened an investigation: Mark Scott, "Cambridge
Analytica Helped 'Cheat' Brexit Vote and US Election, Claims
Whistleblower," Politico, March 27, 2018.

155 On March 19, Facebook hired a digital forensics firm: "Pursuing Forensic
Audits to Investigate Cambridge Analytica Claims," Newsroom post,
March 19, 2018.

155 from a report in the *Guardian*: Harry Davies, "Ted Cruz Using Firm that
Harvested Data on Millions of Unwitting Facebook Users," *Guardian*,
December 11, 2015.

156 In a panel called "In Technology We Trust?": "Salesforce CEO Marc
Benioff: There Will Have to Be More Regulation on Tech from the
Government," video posted on CNBC, January 23, 2018.

156 George Soros delivered a speech: "Remarks Delivered at the World
Economic Forum," George Soros website, January 25, 2018.

157 Wael Ghonim, an Egyptian activist: "Organizer of 'Revolution 2.0'
Wants to Meet Mark Zuckerberg," NBC Bay Area website, February 11,
2011.

157 Sandberg responded defensively: "Sheryl Sandberg Pushes Women to
'Lean In'," *60 Minutes*, CBS, March 10, 2013, can be viewed on YouTube.

159 "People come to a social movement from the bottom up": Maureen Dowd,
"Pompom Girl for Feminism," *New York Times*, February 24, 2013.

161 Dick Durbin and Marco Rubio: Jack Turman, "Lawmakers Call on

Facebook to Testify on Cambridge Analytica Misuse," CBS News online, March 21, 2018.

163 The platform's powerful tracking tools: Julia Angwin and Terry Parris, Jr., "Facebook Lets Advertisers Exclude Users by Race," ProPublica, October 28, 2016.

163 An advertiser could target users by: Natasha Singer, "What You Don't Know about How Facebook Uses Your Data," *New York Times*, April 11, 2018.

164 he never saw "a single audit of a developer": Sandy Parakilas, "Opinion: I Worked at Facebook. I Know How Cambridge Analytica Could Have Happened," *Washington Post*, March 20, 2018.

164 Kaplan had built a formidable DC team: https://www.opensecrets.org /federal-lobbying/clients/lobbyists?cycle=2018&id=D000033563.

166 "Lawmakers seem confused about what Facebook does": Emily Stewart, "Lawmakers Seem Confused about What Facebook Does—and How to Fix It," Vox, April 10, 2018.

166 The hearings would become fodder: Laura Bradley, "Was Mark Zuckerberg's Senate Hearing the 'Worst Punishment of All?'," *Vanity Fair*, April 11, 2018.

167 Over ten hours of testimony: Zach Wichter, "2 Days, 10 Hours, 600 Questions: What Happened When Mark Zuckerberg Went to Washington," *New York Times*, April 12, 2018.

167 Zuckerberg had added three billion dollars: Natasha Bach, "Mark Zuckerberg's Net Worth Skyrocketed $3 Billion during His Senate Testimony and Could Rise Again Today," *Fortune*, April 11, 2018.

Chapter 9: Think Before You Share

170 Social media had played a "determining role": Tom Miles, "U.N. Investigators Cite Facebook Role in Myanmar Crisis," Reuters, March 12, 2018.

170 Sandberg would sit before: *Open Hearing on Foreign Influence Operations' Use of Social Media Platforms (Company Witnesses): Hearing before the U.S. Senate Select Committee on Intelligence*, September 5, 2018. Video can be viewed on https://www.intelligence.senate.gov/ website.

172 Cell phone costs plummeted: Michael Tan, "Tech Makes Inroads in Myanmar," CNET, July 6, 2014.

172 Ashin Wirathu, a Buddhist monk whose anti-Muslim positions: Shashank Bengali, "Monk Dubbed 'Buddhist Bin Laden' Targets Myanmar's Persecuted Muslims," *Los Angeles Times*, May 24, 2015.

174 The post was titled: Mark Zuckerberg, "Is Connectivity a Human Right," Facebook blog post, August 21, 2013.

176 meeting with President Xi Jinping twice in 2015: Jane Perlez, "A Chat in

Chinese with Mark Zuckerberg, as Tech Giants Jostle for Face Time," *New York Times*, September 24, 2015.

177   "Whenever any technology or innovation comes along": Lev Grossman, "Inside Facebook's Plan to Wire the World," *Time*, December 15, 2014.

179   Two people were killed, and fourteen were injured: Tim Hume, "Curfew Imposed after Deadly Clashes between Buddhists, Muslims in Myanmar, CNN website, July 6, 2014.

183   "Facebook Treats You Like a Lab Rat": David Goldman, "Facebook Treats You Like a Lab Rat," CNN website, June 30, 2014.

185   In June 2015, Facebook announced a change to the News Feed: Anisha Yu and Sami Tas, "Taking Into Account Time Spent on Stories," Facebook blog post, June 12, 2015.

186   Facebook took down the posts: Paul Mozur, "A Genocide Incited on Facebook, With Posts from Myanmar's Military," *New York Times*, October 15, 2018.

186   Independent Investigative Mechanism for Myanmar: United Nations General Assembly, "Resolution Adopted by the Human Rights Council on 27 September 2018," https://undocs.org/en/A/HRC/RES/39/2.

Chapter 10: The Wartime Leader

190   She reminded everyone of the company's mission statement: Josh Constine, "Facebook Changes Mission Statement to 'Bring the World Closer Together,'" TechCrunch, June 22, 2017.

192   "Peacetime CEO works to minimize": Ben Horowitz, "Peacetime CEO/ Wartime CEO," blog post on a16z website, April 14, 2011.

192   In public speeches and interviews: Ben Horowitz, "Ben Horowitz on the Lessons He Learned from Intel's Andy Grove," March 22, 2016, Bloomberg interview can be accessed on Youtube.

193   came on the heels of a major reorganization: Kurt Wagner, "Facebook Is Making its Biggest Executive Shuffle in Company History," Vox, May 8, 2018.

194   Jan Koum, the cofounder of WhatsApp: Elizabeth Dwoskin, "WhatsApp Founder Plans to Leave after Broad Clashes with Parent Facebook," *Washington Post*, April 30, 2018.

194   Zuckerberg had broken his promise: *State of New York et al. v. Facebook*.

194   Kevin Systrom, the cofounder of Instagram: Sarah Frier, "Instagram Founders Depart Facebook after Clashes with Zuckerberg," *Bloomberg*, September 24, 2018.

195   It was "really cool for identifying acquisition targets": *State of New York et al. v. Facebook*.

200 Seated just one row back: Mike Isaac, "Rifts Break Open at Facebook over Kavanaugh Hearing," *New York Times*, October 4, 2018.

202 The figures on diversity at Facebook were only inching forward: Maxine Williams, "Facebook 2018 Diversity Report: Reflecting on Our Journey," Facebook blog post, July 12, 2018.

203 As Swisher grilled him about the controversy: Dan Nosowitz, "Mark Zuckerberg Gives Awkward, Sweaty Interview at D8: Touches on Privacy and Scandal," *Fast Company*, June 3, 2010.

204 Alex Jones and his conspiracy-laden site: Hadley Freeman, "Sandy Hook Father Leonard Pozner on Death Threats: 'I Never Imagined I'd Have to Fight for My Child's Legacy,'" *Guardian*, May 2, 2017.

204 Swisher pressed Zuckerberg: Kara Swisher, "Facebook CEO Mark Zuckerberg on Recode Decode," July 18, 2018, podcast interview and transcript can be accessed on the Vox website.

206 Zuckerberg tried to clarify his comments: Kara Swisher, "Mark Zuckerberg Clarifies: 'I Personally Find Holocaust Denial Deply Offensive, and I Absolutely Didn't Intend to Defend the Intent of People Who Deny That'," Vox, July 18, 2018.

207 He had floated the idea of an outside panel: Casey Newton, "Facebook Wants a Social Media Supreme Court So It Can Avoid Hard Questions," The Verge, April 3, 2018.

209 election interference as an "arms race": "Removing Bad Actors on Facebook," Facebook blog post, July 31, 2018.

209 as well as a campaign of hundreds: Sheera Frenkel and Nicholas Fandos, "Facebook Identifies New Influence Operations Spanning Globe," *New York Times*, August 21, 2018.

215 On November 14, 2018, the piece: Sheera Frenkel, Nicholas Confessore, Cecilia Kang, Matthew Rosenberg, and Jack Nicas, "Delay, Deny and Deflect: How Facebook's Leaders Fought through Crisis," *New York Times*, November 14, 2018.

Chapter 11: Coalition of the Willing

219 An op-ed published in the *New York Times*: Chris Hughes, "Opinion: It's Time to Break up Facebook," *New York Times*, May 9, 2019.

221 When a reporter for France 2 television news: Zuckerberg's May 10, 2019 interview with Laurent Delahousse of France 2 television can be viewed on the francetvinfo website.

222 "If what you care about": Ibid.

225 the FTC was poised to fine the company: Lauren Feiner, "Facebook Says the FTC Privacy Inquiry Could Cost as Much as $5 Billion," CNBC website, April 24, 2019.

227 the plan to break down the walls: Mike Isaac, "Zuckerberg Plans to

Integrate WhatsApp, Instagram and Facebook Messenger," *New York Times*, January 25, 2019.

227 Making the apps "interoperable": Mark Zuckerberg, "A Privacy-Focused Vision for Social Networking," Facebook blog post, March 6, 2019.

227 Blue, Instagram, and WhatsApp: U.S. House Subcommittee on Antitrust, Commercial and Administrative Law of the Committee of the Judiciary, *Investigation of Competition in Digital Markets*, October 6, 2020, page 136.

227 2.6 billion users globally: Isaac, "Zuckerberg Plans to Integrate WhatsApp, Instagram and Facebook Messenger."

228 After the merger, Instagram had stayed at arm's length: Mike Isaac, "When Zuckerberg Asserted Control, Instagram's Founders Chafed," *New York Times*, September 25, 2018.

228 Zuckerberg, who feared Instagram: Sarah Frier, *NO FILTER: The Inside Story of Instagram* (New York: Simon and Schuster, 2020), pp. 226–28.

228 Instagram introduced its first ads in November 2013: Vindu Goel and Sydney Ember, "Instagram to Open Its Photo Feed to Ads," *New York Times*, June 2, 2015.

228 an estimated value of $100 billion: Emily McCormick, "Instagram Is Estimated to Be Worth More than $100 Billion," *Bloomberg*, June 25, 2018.

229 Zuckerberg and Sandberg's involvement: Sarah Frier, "Zuckerberg's Jealousy Held Back Instagram and Drove off Founders: An Excerpt from *No Filter*," *Bloomberg*, April 7, 2020.

229 The story was similar with WhatsApp: Parmy Olson, "WhatsApp Cofounder Brian Acton Gives the Inside Story on #DeleteFacebook and Why He Left $850 Million Behind," *Forbes*, September 26, 2018.

231 "The files show evidence of Facebook taking aggressive positions": Isobel Asher Hamilton, "Emails Show Mark Zuckerberg Personally Approved Facebook's Decision to Cut off Vine's Access to Data," Business Insider, December 5, 2018.

Chapter 12: Existential Threat

233 Sandberg met with House Speaker Nancy Pelosi: Emily Birnbaum, "Facebook COO Sheryl Sandberg Meets with Senators on Privacy," The Hill, May 7, 2019.

233 "What a pleasure to visit": Nancy Pelosi, Facebook post, July 22, 2015.

234 On Wednesday, May 22, 2019: Drew Harwell, "Faked Peolosi Videos, Slowed to Make Her Appear Drunk, Spread across Social Media, *Washington Post*, May 23, 2019.

235 It racked up more than two thousand comments: Ed Mazza, "WHOOPS: Giuliani Busted with Doctored Pelosi Video as He Tweets about Integrity," Huffington Post, May 23, 2019.

235 She doesn't drink alcohol: Sarah Mervosh, "Distorted Videos of Nancy Pelosi Spread on Facebook and Twitter, Helped by Trump," *New York Times,* May 24, 2019.

235 In the original broadcast: The May 22, 2019 "Speaker Pelosi at CAP Ideas Conference" video can be viewed on C-Span's website.

236 Within hours, YouTube had removed the clip: Brian Fung, "Why It Took Facebook So Long to Act against the Doctored Pelosi Video," CNN website, May 25, 2019.

236 Rep. David Cicilline of Rhode Island tweeted: David Cicilline, "Hey @ facebook, you are screwing up," tweet posted May 24, 2019.

236 Sen. Brian Schatz called out the platform: Brian Schatz, "Facebook is very responsive to my office when I want to talk about federal legislation," tweet posted May 24, 2019.

237 Macron and Jacinda Ardern: Ryan Browne, "New Zealand and France Unveil Plans to Tackle Online Extremism without the US on Board," CNBC website, May 15, 2019.

239 Weeks later, during a panel: "Sheryl Sandberg Talks Diversity and Privacy at Cannes Lions," June 19, 2019, video can be viewed on Facebook.

240 "Chopping a great American success story into bits": Nick Clegg, "Breaking up Facebook Is Not the Answer," *New York Times*, May 11, 2019.

241 in an internal post in March: Ryan Mac, "Mark Zuckerberg Tried Hard to Get Facebook into China. Now the Company May be Backing Away," Buzzfeed News, March 6, 2019.

241 The Federal Trade Commission opened an investigation: Cecilia Kang, David Streitfeld, and Annie Karni, "Antitrust Troubles Snowball for Tech Giants as Lawmakers Join In," *New York Times*, June 3, 2019.

241 Led by New York, eight state attorneys general: John D. McKinnon, "States Prepare to Launch Investigations into Tech Giants," *Wall Street Journal*, June 7, 2019.

241 The House Judiciary Committee's antitrust subcommittee began a separate inquiry: Cecilia Kang, "House Opens Tech Antitrust Inquiry with Look at Threat to News Media," *New York Times*, June 11, 2019.

242 In two Q&A meetings in July: Casey Newton, "All Hands on Deck," The Verge, October 1, 2019.

Chapter 13: The Oval Interference

243 Behind Zuckerberg, a small gold statue of Poseidon: Chip Somodevilla, "President Donald Trump Welcomes NATO Secretary General Jens Stoltenberg to the White House," www.getty.images.com, News Collection #1139968795.

244 "Facebook was always anti-Trump": Maya Kosoff, "Trump Slams

Zuckerberg: 'Facebook Was Always Anti-Trump'," *Vanity Fair*, September 27, 2017.

246 dinners and meetings for Zuckerberg with influential conservatives: Natasha Bertrand and Daniel Lippman, "Inside Mark Zuckerberg's Private Meetings with Conservative Pundits," Politico, October 14, 2019.

247 From the fourteenth floor of an office tower: Don Alexander Hawkins, "Welcome to Rosslyn, Team Trump. Here's All You Need to Know," Politico, December 16, 2018.

247 They were planning to spend at least $100 million: Grace Manthey, "Presidential Campaigns Set New Records for Social Media Ad Spending," ABC7 News online, October 29, 2020.

247 more than double the amount from the 2016 campaign: Bryan Clark, "Facebook Confirms: Donald Trumped Hillary on the Social Network during 2016 Election," TNW, April 4, 2018.

247 she traveled to Atlanta as the featured speaker: Jeff Amy, "Advocates Fault Facebook over Misleading Posts by Politicos," Associated Press, September 26, 2019.

248 On June 30, 2019, she had: "Facebook's Civil Rights Audit Progress Report," June 30, 2019. PDF can be accessed on FB website.

249 Clegg dropped a bombshell: Nick Clegg, "Facebook, Elections and Political Speech," Facebook blog post, September 24, 2019.

250 When she took the stage: A.R. Shaw, "Facebook's Sheryl Sandberg Confronts Race, Diversity at 'Civil Rights x Tech'," A.R. Shaw, Rolling Out, October 4, 2019.

255 Sherrilyn Ifill, president of the NAACP's Legal Defense Fund: Sherrilyn Ifill, "Opinion: Mark Zuckerberg Doesn't Know His Civil Rights History," *Washington Post*, October 7, 2019.

256 "Free speech and paid speech": "Read the Letter Facebook Employees Sent to Mark Zuckerberg about Political Ads," *New York Times*, October 28, 2019.

256 During his trip, he met Trump again: Ben Smith, "What's Facebook's Deal with Donald Trump?" *New York Times*, June 21, 2020.

257 A survey conducted by the Factual Democracy Project: Andrea Germanos, "Poll Shows Facebook Popularity Tanking. And People Don't Like Zuckerberg Much Either," Common Dreams, March 30, 2018.

261 Every January, for more than a decade: Mary Meisenzahl and Julie Bort, "From Wearing a Tie Every Day to Killing His Own Meat, Facebook CEO Mark Zuckerberg Has Used New Year's Resolution to Improve Himself Each Year," Business Insider, January 9, 2020.

261 "protecting our community": Mark Zuckerberg, Facebook post, January 4, 2018.

Chapter 14: Good for the World

265 COVID-19 virus was spreading: "WHO/Coronavirus International Emergency," January 30, 2020 video can be viewed on UNifeed website.

270 seemed to criticize Twitter's approach: Yael Halon, "Zuckerberg Knocks Twitter for Fact-Checking Trump," Fox News online, May 27, 2020.

272 Three days later thirty-three former early employees: https://assets .documentcloud.org/documents/6936057/Facebook-Letter.pdf.

274 He had spoken openly about his belief: Kif Leswing, "Top Facebook Exec Who Left this Year Says Political Ads Should Be Fact-checked," CNBC online, November 8, 2019.

274 "Social media companies and messaging companies": "Fireside Chat with Chris Cox, Former CPO of Facebook," July 16, 2019, can be viewed on Youtube.

275 an initiative spearheaded by a broad group: Kim Lyons, "Coca-Cola, Microsoft, Starbucks, Target, Unilever, Verizon: All the Companies Pulling Ads from Facebook," The Verge, July 1, 2020.

279 "I fully plan to kill looters and rioters tonight": Ryan Mac and Craig Silverman, "How Facebook Failed Kenosha," Buzzfeed News, September 3, 2020.

280 Facebook removed the event: Ibid.

283 In September alone, the president: Kevin Roose, "Trump's Covid-19 Scare Propels Him to Record Facebook Engagement," *New York Times*, October 8, 2020.

287 "I think these events were largely organized": Reuters, "An Interview with Facebook's Sheryl Sandberg," January 11, 2021, video can be viewed on Youtube.

290 User reports of violent content: Jeff Horwitz, "Facebook Knew Calls for Violence Plagued 'Groups,' Now Plans Overhaul, *Wall Street Journal*, January 31, 2021.

Epilogue: The Long Game

295 "performative changes," as Sen. Elizabeth Warren described: Elizabeth Warren, "Facebook Is Again Making Performative Changes to Try to Avoid Blame for Misinformation in its platform," Facebook post, October 7, 2020.

295 "A 'breakup' of Facebook is thus a complete nonstarter": Jeff Horwitz, "Facebook Says Government Breakup of Instagram, Whatsapp Would Be 'Complete Nonstarter'," *Wall Street Journal*, October 4, 2021.

298 On January 27, 2021: "FB Q4 2020 Earnings Call Transcript," January 28, 2021, Motley Fool website.

300 Libra blockchain currency project: Diem, "Announcing the Name Diem," press release dated December 1, 2020, can be found on Diem.com.

Index

About the Authors

Sheera Frenkel covers cybersecurity from San Francisco for the *New York Times*. Previously, she spent more than a decade in the Middle East as a foreign correspondent, reporting for BuzzFeed, NPR, the *Times* of London, and McClatchy newspapers.

Based in Washington, DC, **Cecilia Kang** covers technology and regulatory policy for the *New York Times*. Before joining the paper in 2015, she reported on technology and business for the *Washington Post* for ten years.

Frenkel and Kang were part of the team of investigative journalists recognized as 2019 Finalists for the Pulitzer Prize for National Reporting. The team also won the George Polk Award for National Reporting and the Gerald Loeb Award for Investigative Reporting.